Statistical Mechanics and its Chemical Applications

Statistical Mechanics and its Chemical Applications

M. H. EVERDELL

Department of Chemistry,
University of Aston in Birmingham,
Birmingham, England.

 1975

ACADEMIC PRESS
London · New York · San Francisco
A Subsidiary of Harcourt Brace Jovanovich, Publishers

ACADEMIC PRESS INC. (LONDON) LTD
24–28 Oval Road
London NW1 7DX

U.S. Edition published by
ACADEMIC PRESS INC.
111 Fifth Avenue
New York, New York 10003

Library of Congress Catalog Card Number: 75 19634

ISBN: 0 12 244450 7

Filmset by Technical Filmsetters Europe Ltd., Manchester
and printed by J. W. Arrowsmith Ltd., Bristol

Preface

Statistical mechanics is that subject which seeks to provide an explanation for the behaviour of macroscopic systems in terms of the characteristics of their constituent molecules. As such it is an indispensable part of chemistry. This is an introductory text which explains the foundations of the subject and some of its chemical applications. It is intended primarily for under-graduate study, (although most undergraduates would not be expected to make a detailed study of all sections), but I hope that it will also be found to provide useful and stimulating reading for chemists of all ages. It assumes that the reader has taken a course in classical thermodynamics. The quantum mechanics required for an understanding of the text has been kept to a minimum: in fact all that is really required is a knowledge of quantum-mechanical concepts such as energy levels and their degeneracies. Although much of statistical mechanics appears to consist of the manipulation of mathematical formulae, the mathematical techniques used in developing the subject are few, (the theorems required are listed on pages x, xi and xii), and in the main are covered by elementary courses in differential calculus and combinatorial mathematics.

There are essentially two ways of approaching statistical mechanics, one based on a study of independent particles in an isolated system, and the other based on studies of the canonical and grand canonical ensembles; and the most important question that the writer of an introductory text has to decide is which approach to choose. The question is finely balanced. The conceptual foundations of the first approach are simple to understand, and lead directly to some results which are seen to be relevant to chemistry; but this approach has two disadvantages. The first is that its relations are applicable only to systems of independent particles, and therefore cannot be used to throw light on some topics of greatest interest to chemists, and the second is that to use its methods in the investigation of many topics in physical chemistry (even though these are concerned only with independent particles) proves to be tedious and often rather involved. On the other hand, the ensemble approach is difficult (unless developed as an elaboration of the

v

first, *and* presented only after the first is clearly understood), and takes longer
to reach results which are seen to be relevant, but when once mastered
provides relations which may be applied to 'advanced' topics (even those
concerned with systems of interacting particles) with comparative ease.

I have attempted to secure the advantages of both approaches. The first
part of the book describes the statistical mechanics of isolated systems of
independent particles, shows how the properties of atomic crystals and
perfect gas systems may be calculated from the values of molecular parameters obtained via quantum theory from spectroscopic data, and how these
may be used in the evaluation of equilibrium constants for reactions of
perfect gases. This covers all the material which some university teachers
would consider an *essential* part of a first degree chemistry course. In the
second part of the book are presented the mechanics of the canonical and
grand canonical ensembles, (the treatment of which may now be presented
with comparative ease because of the familiarity with the methods of statistical
mechanics acquired in Part I), and the use of their relations in the elucidation
of some important topics in physical chemistry.

The first part of the book is addressed to those chemists who wish to use
statistical mechanics only to answer such questions as why a hard element
of low atomic weight fails to obey the Dulong and Petit law, why because the
molecular structure of one isomer differs from that of another one is more
readily obtained than the other, why the equilibrium constant for one
reaction is small and that of another large.

The second part of the book is addressed to those who wish to know
why many of the findings of physical chemistry are as they are: the molecular
origin of the adsorption isotherms, the reason why the equations of one are
obeyed in some circumstances but not others; the molecular origin of
Raoult's law, why some liquid mixtures which fail to obey it exhibit positive
deviations and some negative, why some of only the former type exhibit
limited miscibility, and so on. All questions such as these are answered by
the mechanics of the canonical ensemble, but for those readers who wish to
explore the even more elegant techniques of the grand canonical ensemble,
two chapters are devoted to the construction of its relations and to some of
their uses.

The book is so constructed that much of Part II is accessible to the
reader who does not wish to construct the relations of the ensembles. Thus
the derivation of the condition for phase equilibrium using only the methods
described in Part I, is the subject of Exercise 9,2, and opens the way to the
remainder of Chapter 9. Similarly the foundation of the Langmuir adsorption
isotherm using only the earlier methods is the subject of Exercise 10,3 and
opens the way to all of value in Chapters 10 and 13. Only the statistical-
mechanical interpretation of the chemistry of liquid mixtures is accessible
only to the reader who has mastered the methods of the canonical ensemble.

The book is also so constructed that the reader may if he so wishes, use the relations of the canonical ensemble throughout the greater part of the book. In this event he is recommended, first to study Chapters 1 and 2, and then to pass directly to Chapter 8. He may then proceed

   (a) to Chapter 3 via equation (8,78),
   (b) to Section 4,4 and Chapter 5 via equation (8,84),
   (c) to Section 6,4 via equation (8,85),
and (d) to obtain the condition for chemical equilibrium via Exercise 8,3, and then pass directly to Section 7,2 and beyond.

A few remarks must now be made regarding the topics omitted from this book. The inclusion of a section on classical (pre-quantum theory) statistics is, in my opinion, unnecessary. All that I gained from a superficial study of this topic was an enhanced respect for those who, in pre-quantum theory days laid the foundations of statistical mechanics. Neither has it seemed desirable in an introductory text to discuss the relations of Bose–Einstein and Fermi–Dirac quantum statistics, other than brief references in Sections 4,2 and 5,7. I was thus precluded from discussing such topics as the rôle of electrons in a metallic conductor and the properties of helium at low temperatures. I have made no mention of the statistical-mechanical origin of the so-called theory of absolute reaction rates. Treatments of this subject are best left to books concerned with reaction kinetics. It would have been pleasant to have included a chapter on the apparently most attractive use of the relations of the grand canonical ensemble to throw light on the origin of the virial coefficients of an imperfect gas, but I fail to follow a method which appears to depend on the expansion of an expression $\ln(1 + x)$ where x is very much greater than unity. The fault is undoubtedly mine. Those who are untroubled by such an objection are referred to lucid accounts of this topic in the books listed as 3 and 4 in Appendix III.

It cannot be emphasised too strongly that the exercises given at the end of most chapters should be regarded as an integral part of the text, and their completion is recommended to all readers. Some may appear to be only arithmetical exercises requiring only the substitution of figures into a complicated formula and the evaluation of the result. They should not be despised. Better chemists than I have failed to obtain a sensible value for an equilibrium constant through expressing a standard pressure in inappropriate units.

In conclusion I must thank the Senate of the University of Aston for a period of study leave in which to complete this book, and those of my colleagues whose duties were re-arranged so as to make my leave possible.

University of Aston,   M. H. EVERDELL
Birmingham.

Contents

Part I

The Statistical Mechanics of Isolated systems of Independent Particles

Chapter 1. The Fundamental Principles of Statistical Mechanics

Chapter 6.   Perfect Gas Mixtures

Chapter 7.   The Principles of Chemical Equilibrium in Perfect Gas Reactions

Part I I

The Statistical Mechanics of the Canonical and Grand Canonical Ensembles

Chapter 11.   Liquid Mixtures

Chapter 12.   The Grand Canonical Ensemble

Chapter 13.   The Adsorption of Gases Re-Visited

Contents

Appendices

Mathematical Theorems Used in this Book

The mathematical theorems used in this book are limited to the following:

1. For $-1 < x < 1$,

$$1 + x + x^2 + x^3 + \cdots + x^n + \cdots = \frac{1}{1-x}.$$

2. For all values of x,

$$\exp x = 1 + x + \frac{x^2}{2!} + \frac{x^3}{3!} + \cdots + \frac{x^n}{n!} + \cdots .$$

3. The binomial theorem,

$$(1 + x)^B = \sum_{n=0}^{n=B} \frac{B!}{n!\,(B-n)!} x^n.$$

4. The multinomial theorem,

$$(1 + x + y)^B = \sum_{n=0,m=0}^{n+m=B} \frac{B!}{n!\,m!\,(B-n-m)!} x^n y^m.$$

5. $\dfrac{d}{dx} x^n = nx^{n-1}.$

6. $\dfrac{d}{dx} \ln x = \dfrac{1}{x}$, and if $f(x)$ is any function of x,

$$\frac{d}{dx}\ln f(x) = \frac{1}{f(x)}\frac{d}{dx}f(x).$$

7. $\dfrac{d}{dx}\exp x = \exp x$, and if $f(x)$ is any function of x,

$$\frac{d}{dx}\exp f(x) = \exp f(x)\frac{d}{dx}f(x).$$

8. If C is a function of $X, Y, Z, \ldots$ simultaneous changes in $X, Y, Z, \ldots$ result in a total change in C given by the expression

$$dC = \left(\frac{\partial C}{\partial X}\right)_{Y,Z,\ldots} dX + \left(\frac{\partial C}{\partial Y}\right)_{X,Z,\ldots} dY + \left(\frac{\partial C}{\partial Z}\right)_{X,Y,\ldots} dZ + \cdots.$$

Two extensions of this expression are used in Chapter 12:

(i) If C is a function of X and Y, and X and Y are functions of E and F,

$$\left(\frac{\partial C}{\partial E}\right)_F = \left(\frac{\partial C}{\partial X}\right)_Y\left(\frac{\partial X}{\partial E}\right)_F + \left(\frac{\partial C}{\partial Y}\right)_X\left(\frac{\partial Y}{\partial E}\right)_F, \quad \text{and}$$

(ii) If C is a function of X, Y and Z and X, Y and Z are functions of E and F,

$$\left(\frac{\partial C}{\partial E}\right)_F = \left(\frac{\partial C}{\partial X}\right)_{Y,Z}\left(\frac{\partial X}{\partial E}\right)_F + \left(\frac{\partial C}{\partial Y}\right)_{X,Z}\left(\frac{\partial Y}{\partial E}\right)_F + \left(\frac{\partial C}{\partial Z}\right)_{X,Y}\left(\frac{\partial Z}{\partial E}\right)_F.$$

9. We use only one special integral:

$$\int_{n_x=1}^{n_x=\infty} \exp\left[-\frac{An_x^2}{x^2}\right] dn_x = \frac{1}{2}\left[\frac{\pi x^2}{A}\right]^{\frac{1}{2}}.$$

10. Lagrange's method of undetermined multipliers:
    The substance of this method, which is used repeatedly throughout the book, is discussed in detail on pages 42–44.

11. The number of ways in which N *distinguishable* objects may be assigned to G labelled containers (there being no restriction on the number of objects in each) is G^N.

12. The number of ways in which N *distinguishable* objects may be assigned to G labelled containers so that there are $N_1, N_2, \ldots N_r, \ldots$ objects in each is

$$\frac{N!}{N_1!\,N_2!\cdots N_r!\cdots} = \frac{N}{\prod_r N_r!}$$

where

$$\sum_r N_r = N$$

13. The number of ways in which N *indistinguishable* objects may be assigned to G labelled containers, (there being no restriction on the number of objects in each) is

$$\frac{(G + N - 1)!}{(G - 1)! \, N!}.$$

14. The number of ways in which N *indistinguishable* objects may be assigned to G labelled containers, $(G > N)$, so that each is occupied by one object or none, is

$$\frac{G!}{N!(G - N)!}.$$

15. The number of distinguishable arrangements of N identical objects of one kind and M identical objects of another kind on $N + M$ sites is

$$\frac{(N + M)!}{N! \, M!}.$$

16. The symbol ! figuring above means the factorial of the number preceding it. Thus $N!$ means factorial N, and is the product

$$N(N - 1)(N - 2) \cdots 3 \cdot 2 \cdot 1$$

The evaluation of factorials of large numbers in expressions such as those given above is accomplished by means of Stirling's approximation formulae:

(i) for values of N greater than twenty, a very accurate value for the natural logarithm of $N!$ is given by the formula

$$\ln N! \approx N \ln N - N + \tfrac{1}{2} \ln (2\pi N).$$

(ii) for values of N greater than one hundred, the simpler formula

$$\ln N \approx N \ln N - N$$

may be used without sensible error.

In this context the symbol $\approx$ should be read as "may be replaced by".

Symbols

A	Helmholtz function $\equiv E - TS$
B	Number of equivalent sites on which adsorption may occur
$^\circ C$	Degree Celsius
C_p	Heat capacity at constant pressure
C_v	Heat capacity at constant volume
D_o	Dissociation energy of a molecule
d	differential coefficient
∂	partial differential coefficient
E	Energy
E^0	Molar energy of pure substance at absolute zero
f	Partition function (for independent particle)
f_e	Electronic partition function
f_{odo}	Partition function for one-dimensional oscillator
f_r	Rotational partition function
f_t	Translational partition function
f_{tdo}	Partition function for three-dimensional oscillator
f_v or f_{vib}	Vibrational partition function
$f' = f/V$	
$f_t' = f_t/V$	
G	Gibbs function $\equiv E + PV - TS \equiv H - TS \equiv A + PV$
g	Degeneracy of an energy level accessible to a particle
H	Enthalpy $\equiv E + PV$
h	Planck constant
I	Moment of inertia
i	Vibrational quantum number for a localised particle
J	Joule $\equiv 1$ Nm
J	Rotational quantum number
K	degrees on the thermodynamic scale of temperature
K	Equilibrium constant, $K_N, K_x, K_{P/P}, \ldots$ equilibrium constants expressed in terms of numbers of molecules, mole fractions, partial pressures . . . .
k	Boltzmann constant
L	Avogadro constant
M	Molar mass
m	Mass of particle
m	metre
$N, M, \ldots$	usually used to denote numbers of particles in system
N	Newton
P	Pressure
P_A	Partial pressure of component A

p Chance, as in $p_{(E_i)}$, the chance that a system in an ensemble acquires energy E_i

Q Canonical partition function

q Quantity of heat absorbed by a system

R Gas constant $\equiv Lk$

S Entropy

s Second

T Temperature

V Volume

v Vibrational quantum number for a non-localised particle

W_{max} Number of complexions encompassed by the most probable configuration of a system of prescribed energy

W_{total} Total number of complexions accessible to a system of prescribed energy

$W_{(E_i)}$ Number of complexions characterised by energy E_i

w Work performed on a system

w, w' Interaction parameters used in the theory of liquid mixtures

x Mole fraction

Z_G Grand partition function

z Number of nearest neighbours (theory of liquid mixtures)

Script letters

$\mathscr{A}$ Area

$\mathscr{E}$ The energy of an ensemble

$\mathscr{N}$ The total number of systems in an ensemble

$\mathscr{N}_i$ The number of systems in a canonical ensemble characterised by energy E_i

$\mathscr{N}_{N,M,j}$ The number of systems in a grand canonical ensemble containing N molecules of one species, M of another and in the jth quantum state for these numbers of moles

$\mathscr{V}$ The volume of an ensemble

Greek letters

α Lagrangian multiplier,
Cubic coefficient of thermal expansion

β Lagrangian multiplier $\equiv \dfrac{1}{kT}$,
Isothermal compressibility coefficient

γ Activity coefficient
δ Very small increment
Δ Finite increment
$\sum_i$ Sum of all terms following for all values of i
ϵ Energy level of a single particle
θ Fraction of total number of sites occupied in an adsorption process
$\theta, \theta_v, \theta_r$ Characteristic temperatures for vibration and rotation
ξ Extent of reaction
λ force constant,
     de Broglie wave-length,
λ Absolute activity
$\ln \lambda$ A Lagrangian multiplier
ν Frequency $= \dfrac{1}{2\pi}\left(\dfrac{\lambda}{m}\right)^{\frac{1}{2}}$
μ_A (Molecular) chemical potential of component A
μ_A^* (Molar) chemical potential of component A
$\mu_A^\dagger$ (Molecular) chemical potential of component A at chosen standard pressure
μ_A° Standard chemical potential of a component of a liquid mixture
π Ratio of circumference of circle to its diameter
$\prod_i$ Running product of all terms for all values of i
σ Symmetry number
ϖ Alternative symbol for the degeneracy of an energy level accessible to an independent particle
ϖ Alternative symbol for chance
ϖ Wave number
Ω Number of quantum states accessible to an ensemble

Subscripts

e evaporation as in ΔH_e,
     electronic as in f_e,
i energy level, as in g_i the degeneracy of energy level i
j alternative symbol to i
max   maximum as in W_{max}
n nuclear as in f_n
p constant pressure as in C_p
r rotational as in f_r
t translational as in f_t
v vibrational as in f_v,
     constant volume as in C_v

Mathematical symbols

$\equiv$ defined by

$=$ equal to

$\simeq$ approximately equal to

$\approx$ may be replaced by

$\neq$ not equal to

$\propto$ proportional to

$<$ less than

$>$ greater than

$\ll$ very much less than

$\gg$ very much greater than

$\rightarrow$ approaches

e exponential i.e. $e^x = \exp x = \sum_{n \geq 0} \dfrac{x^n}{n!}$

$\ln$ logarithm to base e

$\log_{10}$ logarithm to base 10

$!$ factorial i.e. $n! = n(n-1)(n-2)\cdots 3 \cdot 2 \cdot 1.$

$\prod_{i}$ running product of all terms for all values of i

$\sum_{i}$ sum of all terms following for all values of i

Values of Physical Constants

Quantity	Symbol	Value to five significant figures
Boltzmann constant	k	1.3805×10^{-23} JK^{-1}
Gas constant	$R = Lk$	8.3143 JK^{-1} mol^{-1}
Avogadro constant*	L	6.0225×10^{23} mol^{-1}
Faraday constant	Le	9.6487×10^4 C mol^{-1}
Planck constant	h	6.6256×10^{-34} Js
Atomic mass unit	amu	1.6604×10^{-27} kg
mass of electron	m_e	9.1091×10^{-31} kg
electronic charge	e	1.6021×10^{-19} C
electron volt	eV	1.6021×10^{-19} J
Ice-point temperature	T_{ice}	273.15 K
Atmosphere	atm	101.325 kNm^{-2}

* The mole is defined as the quantity of material containing the same number of particles as there are atoms in exactly 0.012 kg of the isotope carbon-12.

Numerical Quantities

Ratio of circumference
of circle to diameter π 3.1416

$$\sum_{N=0}^{N=\infty} \frac{x^N}{N!}$$ $\exp x$ 2.7183^x

$\ln 10$ 2.3026

The Statistical Mechanics of Isolated Systems of Independent Particles

CHAPTER 1

The Fundamental Principles of Statistical Mechanics

1.1 The Relationship between Statistical Mechanics and Classical Thermodynamics

Statistical mechanics may be regarded as a branch of science which is completely independent of classical thermodynamics, or (as it is regarded in this book) as a discipline which is complementary to thermodynamics, in the sense that it throws light on the origin of the thermodynamic laws and, in some cases, provides information unobtainable by the thermodynamic approach. The power of statistical mechanics is best judged by comparison with the power and limitations of thermodynamics.

Classical thermodynamics depends on four postulates (each based on a body of experimental fact) known as the zeroth, first, second and third laws. The zeroth law provides a thermodynamic definition of temperature, the first and second laws establish respectively the significance of the *energy* of a system and its *entropy*, and lay down the rules governing their change, and the third law postulates the equality of the entropies of all pure crystalline substances at absolute zero. The thermodynamic structure based on these postulates unifies a great range of phenomena, and produces a number of relations, some of which are inequalities which enable us to predict the *feasibility* of any projected process, and some of which are equalities which enable us to appreciate the conditions governing the equilibrium state. Typical of the first is the criterion governing the feasibility of a chemical reaction at constant temperature and pressure, the criterion being simply

$$(T\,\Delta S - \Delta H)_{T,P} \geqslant 0 \qquad (1.1)$$

where ΔH is the heat of reaction and ΔS the corresponding entropy change for the system. Typical of the second is the equation giving the effect of

3

pressure on the melting point of a pure solid

$$\frac{dT_f}{dP} = \frac{V_{\text{liquid}} - V_{\text{solid}}}{S_{\text{liquid}} - S_{\text{solid}}} = \frac{\Delta V_f}{\Delta S_f} \qquad (1.2)$$

where ΔV_f is the change in volume resulting from the melting process and ΔS_f the corresponding entropy change.

The *use* of expressions such as these depends on the availability of physical data, heat capacities, heats of reaction, heats of fusion and evaporation . . . and so on. In some cases these quantities are obtainable by direct measurement. In other cases it may be impossible to measure a particular quantity but possible to obtain it indirectly from expressions relating it to others more readily measured. Typical of such expressions is the equation

$$C_v = C_p - \frac{TV\alpha^2}{\beta} \qquad (1.3)$$

which permits the calculation of C_v, the heat capacity of a substance at constant volume, a quantity which is difficult if not impossible to measure directly in the case of a solid or liquid, from C_p the corresponding heat capacity at constant pressure, α the cubic coefficient of thermal expansion, and β the isothermal compressibility coefficient, all of which may be measured directly with ease.

The strength and power of the thermodynamic laws and of such relations as those given above lie in the fact that they apply *to all systems in all circumstances*. But classical thermodynamics imposes certain limitations on itself. The first is that although it tells us *how* a particular situation is governed it does not tell us *why*, or perhaps more fairly, it may give an answer to the question why, *but one phrased only in thermodynamic terms*.

Let us consider a typical problem in chemistry. Suppose that two moles of hydrogen iodide are introduced at 750 K into a vessel of fixed volume and the system isolated so that no interaction with the surroundings is possible. In due course it is found that the temperature and composition of the system change; a mixture of 1.576 moles of hydrogen iodide, 0.212 moles of hydrogen and 0.212 moles of iodine at 700 K being produced, the endothermic reaction $2\text{HI} \rightarrow \text{H}_2 + \text{I}_2$ having proceeded to a particular extent, and no matter how many times the experiment is repeated the result is always the same. Why does the system change its state? Why does it choose to change to this particular state? Let us rephrase the problem using other words.

By the nature of the experiment set up we have prescribed the total energy of the system, its volume, and the number of atoms[1] of each kind (2L hydrogen

---

[1] In statistical mechanics it is more convenient to express the amount of each component by the number of particles rather than by the number of moles.

atoms and $2L$ iodine atoms, L being the Avogadro constant) irrespective of the manner in which they are combined. Let us use the word *configuration* to denote any particular combination of the constituent atoms *identifiable by macroscopic analysis*, and the corresponding division of the total energy between that associated with chemical bonds and that manifested as thermal energy. Thus the initial state of the system represents one configuration characterised by the fact that each hydrogen atom is combined with one iodine and by that division of energy identified by the system's temperature. A very large number of other configurations for the same number of atoms of each kind and the same total energy are obviously possible. One[1] consists of $1.8L$ molecules of HI, $0.1L$ molecules of H_2 and $0.1L$ molecules of I_2 at 725 K, one of $1.2L$ molecules of HI, $0.4L$ molecules of H_2 and $0.4L$ of I_2 at 650 K, one of L molecules of HI, $0.5L$ molecules of H_2 and $0.5L$ of I_2 at 626 K, one of L molecules of H_2 and L molecules of I_2 at 503 K and so on. Other configurations containing atomic hydrogen and iodine are also possible. The system chooses to change spontaneously from its initial configuration to none of these. It chooses instead to change to a configuration corresponding to $1.576L$ molecules of HI, $0.212L$ molecules of H_2 and $0.212L$ of I_2 at 700 K. Why does it choose to change to this configuration rather than to any other one of the very large number available to it? Why does it choose to change at all?

To be sure, thermodynamics produces an answer. It says that this particular configuration is chosen because the values for the entropies of hydrogen, iodine and hydrogen iodide (and the formula for the entropy of mixing) are such that the entropy associated with this particular configuration is greater than that associated with any other commensurate with the same total energy; but if one pursues the matter by asking why the configuration chosen *is* that associated with the maximum possible entropy, the answer given is that this is so because it is in accordance with the second law, (that is to say, our experience in the past is that this is always the case;) and if we then ask why the entropies of hydrogen, iodine and hydrogen iodide have these values the answer given is that these are the values assigned to these substances as the result of measurements of heat capacities, heats of evaporation . . . and so on.

This brings us immediately to the second limitation of thermodynamics, the fact that although it provides many equations such as (1.2) and (1.3) relating a number of quantities, so that if the values of all but one are known that of the last may be calculated, *it is unable to generate from its own equations a value for the last independent of the values of the others*. Thus considering equation (1.2), only if two of the quantities dT_f/dP, ΔV_f and ΔS_f for a particular

---

[1] The reader may care to show, using the information provided in Exercise 1.1 that all the configurations mentioned are indeed states of equal energy.

melting process are determined experimentally may the third be calculated, or considering equation (1.3), only if five of the quantities C_v, C_p, T, V, α and β are known may the sixth be calculated.

The limitations of classical thermodynamics stem from two closely related sources. The first is that thermodynamics is a purely *experimental* branch of science, in the sense that its laws are based on a body of experimental fact, and its equations of little practical use unless values for all but one of the quantities concerned therein are known, and the second is that the quantities with which it is concerned, heat capacities, heats of reaction, temperatures, pressures, ... are obtained from measurements made on the system as a whole, or on some convenient amount of the substances composing it. Neither the measurement of these quantities nor the thermodynamic structure in which they are used is *concerned* with the *nature* of the substances involved, and in neither case is reference made to the individual atoms or molecules involved, so that thermodynamics is able neither to explain its own laws in terms of the constituent particles, nor to interpret its equations to give information on them. In fact, the limitations of thermodynamics stem from the very reason for its strength and power: the fact that its relationships apply to *all* systems whatever their nature.

It is of course reasonable to ask whether the limitations of thermodynamics might not be dissolved by an approach from that area deliberately left without the thermodynamic view-point, the nature of the particles concerned. It is for example, reasonable to ask whether the values of C_p or C_v for say, an assembly of oxygen molecules are available only from measurements made on the *assembly*, and whether it is not possible to compute them if we know the appropriate characteristics of a single oxygen molecule. A quantity of oxygen (at temperatures at which dissociation may be neglected) consists only of a large number of identical molecules, so that it is reasonable to suppose that all the information relating to the behaviour of the assembly must somehow be "contained" in a single molecule, and if we read the molecule aright we should be able to say ... because its characteristics are so and so, the behaviour of the assembly will be such and such ... .

This is precisely the line of attack in statistical mechanics. We view any chemical system as an *assembly of particles* and first of all establish the reason why any assembly (given enough time) appears always to assume a *particular configuration* when many other configurations are available to it. We then consider the *mechanical* characteristics of its constituent particles, (the most important prove to be their masses, their moments of inertia and their fundamental vibrational frequencies) and with the help of quantum theory, determine their possible energy states and modes of behaviour, and with the help of statistics predict the properties of the assembly in its chosen configuration from the possible modes of behaviour of its constituent particles.

It should perhaps be mentioned that the configuration chosen by any assembly depends on the conditions applied. Thus, the particular configuration chosen by a quantity of hydrogen iodide which is isolated from the surroundings is not the same as that chosen if it is maintained at constant temperature, and although it is not profitable to examine fully the significance of this at this stage, some remarks are pertinent. The first is that it is not possible to predict the behaviour of any assembly in equilibrium unless *some* restrictions on its activity are imposed, just as we can calculate the pressure of a gas only if its temperature and the volume it occupies are known. The second is that we have a certain degree of choice over the restrictive conditions. We can choose to assign to the assembly a fixed volume *or* a fixed pressure, a fixed temperature *or* a fixed energy, and either the condition that the total amount of material in the assembly remains constant, or that material may pass into or out of it, (in thermodynamic language whether the system is *closed* or *open*.) Because of this degree of choice, three methods of tackling the problem before us have evolved depending on the conditions chosen. The first method supposes that the assembly is closed and that its volume and energy are fixed, the second that the system is closed and that its volume and temperature are fixed, and the third that the assembly is open but its volume and temperature fixed.

Part I of this book will be concerned with the development and use of the first method, and only in Part II will the second and third methods be introduced. There are good reasons for this. Although the machinery resulting from the first is in some respects less powerful than that resulting from the second or third (and its field of application strictly limited), the conceptual foundations of the first are simpler to grasp, and, in the opinion of the author, the second and third methods are best appreciated only after the first is understood.

1.2 Configurations and Complexions

Our first task is to attempt to produce an explanation for the fact that any macroscopic system, given enough time, appears to choose one particular configuration when many others are accessible to it. It is as well to accept from the start that the reason for this is a purely statistical one, and is a consequence of the fact that any macroscopic physical system is an assembly of a very large number of particles. The rules governing such an assembly are the same whatever the nature of the units concerned, and so can be uncovered by studying the behaviour of any assembly of units we choose. We choose to study the distribution of particles of two kinds among labelled sites. It is true that the model chosen is not a precise analogue of a real chemical system, but it has two points in its favour: one that its analysis is amenable to the simplest calculations, and the other that it reveals the mathematical tools available for the statistical analysis of a real system.

In the preliminary investigation, when we consider the distribution of only four particles of each kind, we can think of them as red balls and black. Later, when we are concerned with much larger numbers we will refer to them as particles of two kinds R and B, if only for the fact that the imagination baulks at the thought of shaking a box containing a million balls.

Figure 1.1 represents a rectangular box divided into two parts (so that we can meaningfully refer to the left-hand side and the right-hand side), the floor of the box being slightly recessed to provide eight cups labelled A to H. Suppose that it contains eight balls, four identical red balls and four identical black, and that the dimensions of balls and box are such that when the box is shaken the balls can pass freely from one side to the other, but that on coming to rest one ball settles in each cup. Suppose too that we are able to subject the box only to macroscopic examination *capable only of determining the number of balls of each kind on each side*. Such examination would reveal five distinct configurations:

configuration 1:   4 red on the left, 4 black on the right,

configuration 2:   3 red and 1 black on the left, 1 red and 3 black on the right,

configuration 3:   2 red and 2 black on the left, 2 red and 2 black on the right,

configuration 4:   1 red and 3 black on the left, 3 red and 1 black on the right,

configuration 5:   4 black on the left, 4 red on the right.

Suppose that the box is shaken and that the balls are subsequently allowed to come to rest, and that the operation is repeated many times. In which configuration would we expect to find the assembly the largest number of times? In other words, which of the five configurations is the most probable?

Let us consider the number of ways in which configuration 2 may arise. It is clear that the situation in which the left-hand side obtains three red balls and one black may arise in four ways, one in which a black falls into cup A, and a red into each of cups B, C and D, one in which a black falls into B and a red into each of cups A, C and D, one in which the black falls into C, and one in which the black falls into D. To each of these four possibilities correspond four ways by which the right-hand side obtains a single red. The number of ways in which the balls on the left may be arranged is independent of the number of possible arrangements on the right, so that the number of arrangements leading to configuration 2 is sixteen, *each of which can be distinguished from the others only if the box is examined by instruments capable of deter-*

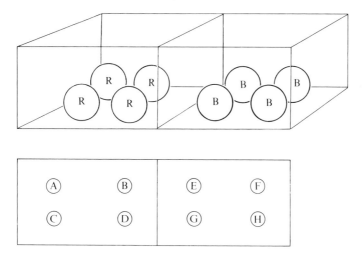

Figure 1.1

mining not only the number of balls of each colour on each side but the position occupied by each of each kind. Let us denote the number of such arrangements leading to a particular configuration by the symbol W, and write

$$W_2 = 16.$$

Using the same reasoning we see that there are six different ways in which two red balls and two black may arise on the left-hand side, (the reds falling in positions AB, AC, AD, BC, BD and CD) and to each of these correspond six ways in which two red and two black arise on the right, so that configuration 3 encompasses thirty six different arrangements. We therefore write

$$W_3 = 36.$$

Since no red (or black) ball is distinguishable from any other of the same colour, only one arrangement leads to configuration 1, so that

$$W_1 = 1,$$

and since configuration 4 is obviously the analogue of configuration 2, and configuration 5 the analogue of configuration 1 we may write

$$W_4 = 16$$

and

$$W_5 = 1.$$

If therefore we inquire the number of distinguishable ways in which four balls of one kind and four of another may be arranged on eight labelled sites in the box described above, we get two different answers depending on our ability to distinguish one arrangement from another. If we employ large-scale instruments capable of telling us only how many balls of each kind are on each side we obtain the answer five, corresponding to configurations 1 to 5, but if we employ more sophisticated instruments capable of telling us the type of ball occupying each site we obtain the answer seventy, $(1 + 16 + 36 + 16 + 1.)$

Whereas we use the word *configuration* to denote those arrangements distinguishable by large-scale instruments (it will be realised that we are using the word in precisely the same sense as it was used in Section 1.1) we use the word *complexion* to denote those arrangements distinguishable only by observation of each particle in turn.

We now come to the point of the exercise. Although the total number of complexions accessible to the system described is seventy, only one of these complexions leads to configuration 1, but sixteen lead to configuration 2, thirty six to configuration 3, sixteen to configuration 4 and one to configuration 5, and *since there would appear to be no prima facie reason why any one complexion should be preferred to any other* (if we shake the box once, the balls are as likely to fall in one way as in any other) there appears to be a chance of only one in seventy that a single fall will result in configuration 1, a chance of sixteen in seventy that it results in configuration 2, a chance of thirty six in seventy that configuration 3 will be reached, and so on, so that if we denote the *chance* that a particular configuration is reached by the symbol ϖ we can write

$$\varpi_1 = 1/70, \varpi_2 = 16/70, \varpi_3 = 36/70, \varpi_4 = 16/70, \varpi_5 = 1/70.$$

These results suggest

(i) that *the chance that any configuration is reached is proportional to the number of complexions encompassed by it*, and is given by the equation

$$\varpi_x = \frac{W_x}{W_{total}} \qquad (1.4)$$

where W_x is the number of complexions leading to configuration X, and W_{total} is the sum of the numbers of complexions encompassed by *all* possible configurations; and

(ii) that one configuration (configuration 3) is more likely to be reached than any other, and so may reasonably be termed the *most probable configuration*.

These conclusions are the crux of the argument on which our approach to statistical mechanics is based, but to be sure of our ground, and particularly to gain further information relating to the most probable configuration we must carry out a more extensive study of the model by considering the distribution of larger numbers of particles.

Since the *ad hoc* method used above to calculate the number of complexions leading to any particular configuration would be impossible for assemblies of large numbers of particles we must first seek a more general method. Combinatorial mathematics tells us that the total number of different ways in which N_R indistinguishable objects of one kind and N_B indistinguishable objects of another can be arranged on $(N_R + N_B)$ sites (i.e. the quantity W_{total} figuring in equation (1.4)) is given by the formula

$$\frac{(N_R + N_B)!}{N_R! \, N_B!}$$

where the symbol ! means the *factorial* of the number preceding it.[1] Since we are here concerned only with the case in which $N_R = N_B$ we may replace both by N to obtain the equation

$$W_{total} = \frac{(2N)!}{N! \, N!}, \tag{1.5}$$

and we observe that when N is four (as in the case discussed above)

$$W_{total} = \frac{8 \cdot 7 \cdot 6 \cdot 5 \cdot 4 \cdot 3 \cdot 2 \cdot 1}{(4 \cdot 3 \cdot 2 \cdot 1)(4 \cdot 3 \cdot 2 \cdot 1)} = 70,$$

as we found before.

We have now to obtain the formula for the number of complexions leading to any particular configuration. Let us consider the configuration in which we have n particles of type R and $N - n$ particles of type B on the left and $N - n$ of type R and n of type B on the right. The number of ways in which the particles on the left may be arranged is (from 1.5) $N!/n! \, (N - n)!$, and the same formula gives the number of possible arrangements of the particles on the right. It follows that the number of complexions leading to this configuration for the system as a whole is given by the equation

$$W_n = \left\{ \frac{N!}{n! \, (N - n)!} \right\}^2 . \tag{1.6}$$

For the case in which $N = 4$ we see that:

when $n = 4$ (configuration 1), $W = 1$,

when $n = 3$ (configuration 2), $W = 16$,

when $n = 2$ (configuration 3), $W = 36$,

when $n = 1$ (configuration 4), $W = 16$,

[1] That is, $n! = n(n - 1)(n - 2)(n - 3) \ldots 3 \cdot 2 \cdot 1$.

and

$$\text{when } n = 0 \text{ (configuration 5),} \qquad W = 1$$

which results are of course the same as those produced before. Using these formulae for the case in which we have six particles of each type we find that the total number of complexions is $(6 + 6)!/6! \, 6!$, i.e. 924, and that of these

400 lead to configuration 3R/3B//3R/3B,

225 to each of configurations 4R/2B//2R/4B and 2R/4B//4R/2B,

36 to each of configurations 5R/1B//1R/5B and 1R/5B//5R/1B

and

1 to each of configurations 6R//6B and 6B//6R.

We see that again a *most probable configuration* has emerged, that corresponding (as before) to equal numbers of particles of each kind on each side,[1] so that we infer that the number of complexions leading to the most probable configuration is obtained by putting $n = N/2$ in (1.6) so giving the equation

$$W_{\text{max}} = \left\{ \frac{N!}{(N/2)! \, (N/2)!} \right\}^2 \tag{1.7}$$

where we use the symbol W_{max} to denote the number of complexions encompassed by the most probable configuration.

The evaluation of expressions such as these appears formidable for large values of N, but in fact the larger the numbers concerned the simpler does the problem become, because for values of N greater than twenty a very accurate value for the natural logarithm of factorial N, $\ln N!$, is given by the *Stirling approximation formula*

$$\ln N! \approx N \ln N - N + \tfrac{1}{2} \ln (2\pi N), \tag{1.8}$$

and for values of N greater than one hundred the simpler formula

$$\ln N! \approx N \ln N - N \tag{1.9}$$

may be used without significant error.

Figures 1.2(i), (ii) and (iii) and Tables 1.1 and 1.2 give some of the results of calculations carried out on the model for larger values of N. Figure 1.2(i) shows pictorially the distribution of complexions among different configurations for the model containing five hundred particles of each kind, the length

---

[1] In the cases discussed in this section the most probable configuration is that corresponding to complete mixing. This is a special feature of this particular model, and is not always the case. We shall find that it is not so when, in later chapters we study systems subject to factors which make *a completely random distribution* unlikely.

of the vertical line through the point 250/250 representing the number of complexions leading to configuration 250R/250B//250R/250B calculated from the equation

$$W_{250/250} = \left\{ \frac{500!}{250!\,250!} \right\}^2, \qquad \text{i.e. } 1.35 \times 10^{298},$$

the length of the vertical line through the point 249/251 representing the number of complexions leading to configuration 249R/251B//251R/249B calculated from the equation

$$W_{249/251} = \left\{ \frac{500!}{249!\,251!} \right\}^2, \qquad \text{i.e. } 1.34 \times 10^{298}$$

and so on, *the factorial in each case being evaluated using expression* (1.8), (the reason for *not* using the simpler expression (1.9) will be apparent later.)

We see (as we expect) that the number of complexions leading to configuration 250R/250B//250R/250B exceeds the number leading to any other, so that this may justifiably be called the most probable configuration. The total number of configurations accessible to the model for five hundred particles of each kind is five hundred and one,[1] but the number of complexions leading to each of the four hundred and fifty configurations less completely mixed than 225R/275B//275R/225B and 275R/225B//225R/275B proves to be too small to be shown on a diagram of the scale to which Figure 1.2(i) is drawn. Thus *almost all complexions are found to be encompassed by relatively few configurations centred around the most probable distribution.*

Figure 1.2(ii) shows pictorially the distribution of complexions among different configurations for the model containing fifty thousand particles of each kind. In this case the total number of possible configurations is fifty thousand and one, but we have shown only five: configuration 25 000R/25 000B//25 000R/25 000B which encompasses $0.81 \times 10^{30\,098}$ complexions, configurations 24 900R/25 100B//25 100R/24 900B and 25 100R/24 900B//24 900R/25 100B each of which encompasses $0.36 \times 10^{30\,098}$ complexions and configurations 24 800R/25 200B//25 200R/24 800B and 25 200R/24 800B//24 800R/25 200B each of which encompasses $0.033 \times 10^{30\,098}$ complexions. (The reader must imagine ninety nine lines between each pair shown.) We surmise from the figure that almost all complexions are encompassed by only a few hundred of the fifty thousand and one possible configurations.

Figure 1.2(iii) gives the results of calculations made on the model containing five hundred thousand particles of each kind. In this case the total

---

[1] Each configuration for this model is identified by the number of particles of one kind on either side. For an assembly of N particles of each kind each side can contain 0, 1, 2, 3 ... $(N-1)$ or N particles of one kind, so that there are $N + 1$ configurations in all.

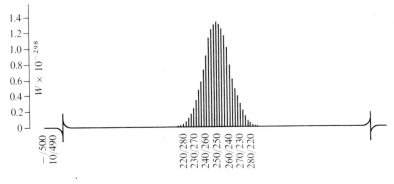

Figure 1.2(i)

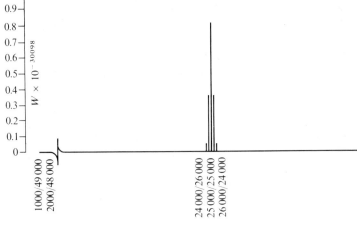

Figure 1.2(ii)

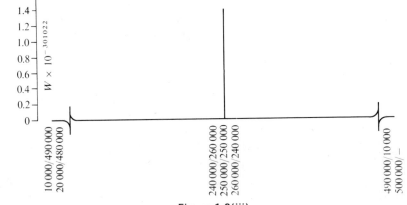

Figure 1.2(iii)

number of possible configurations is five hundred thousand and one, and the number of complexions encompassed by the most probable configuration $250\,000R/250\,000B//250\,000R/250\,000B$ is $1.4 \times 10^{301\,022}$, but the number encompassed by any configuration differing sufficiently from the most probable configuration to be shown on the figure proves to be too small to be shown.

We surmise from these results that the larger the number of particles in the model the smaller is the fraction of the total number of possible configurations encompassing the majority of complexions.

Table 1.1 expresses this conclusion in rather more quantitative terms. It shows for each of four values of N, the three corresponding to Figure 1.2 and an intermediate value, the total number of complexions, P the total number of possible configurations and p the number of configurations encompassing ninety five per cent of all complexions.

Table 1.1

$N = N_R = N_B$	W_{total}	P	p	p/P
500	2.7×10^{299}	501	33	0.066
5\,000	1.6×10^{3008}	5\,001	99	0.0198
50\,000	2.5×10^{30100}	50\,001	311	0.0062
500\,000	5.6×10^{301026}	500\,001	967	0.00193

We see that ninety five per cent of all complexions accessible to the system containing five hundred particles of each kind are encompassed by 6.6 per cent of all configurations, which confirms the qualitative picture given by Figure 1.2(i), but, that as suggested by Figures 1.2(ii) and (iii), this percentage decreases dramatically for larger values of N, so that for the system containing five hundred thousand particles of each kind ninety five per cent of all complexions are encompassed by less than 0.2 per cent of all configurations, the 967 configurations lying between $249\,517R/250\,483B//250\,483R/249\,517B$ and $250\,483R/249\,517B//249\,517R/250\,483B$ and centred around the most probable configuration $250\,000R/250\,000B//250\,000R/250\,000B$.

Now our interest as chemists lies, *not* in the number of ways in which particles may be arranged in a box, but in the behaviour of physical systems, and even the largest number of particles in the assemblies studied above is quite insignificant when compared with the number of molecules contained in the smallest system handled in the laboratory. Macroscopic measurements leading to values for specific heats, latent heats, heats of reaction and so on can hardly be made on less than about one thousandth of a mole of material, and this quantity contains about 10^{20} molecules, so our next task is to use

the results recorded in Table 1.1 to give information on assemblies characterised by values of N of this order. This is quite easily done. Inspection of the
results given in the table shows that for each assembly the value of the fraction
p/P is inversely proportional to $N^{\frac{1}{2}}$, the value of the proportionality constant
being about 1.4. Thus for $N = 500$,

$$p/P = 0.066 \simeq \frac{1.4}{500^{\frac{1}{2}}} \simeq \frac{1.4}{22},$$

for $N = 5000$

$$p/P = 0.0198 \simeq \frac{1.4}{5000^{\frac{1}{2}}} \simeq \frac{1.4}{71}$$

for $N = 50\,000$

$$p/P = 0.0062 \simeq \frac{1.4}{50\,000^{\frac{1}{2}}} \simeq \frac{1.4}{220}$$

and for $N = 500\,000$

$$p/P = 0.00193 \simeq \frac{1.4}{500\,000^{\frac{1}{2}}} \simeq \frac{1.4}{710}.$$

The value of p/P for $N = 10^{20}$ is therefore about 1.4×10^{-10}. Thus for a
model of the type described containing about 10^{20} of each kind, almost all
complexions are encompassed by only about 1×10^{-8} per cent of all possible
configurations, those lying between configuration

$(0.5 \times 10^{20} - 0.5 \times 10^{10})R/(0.5 \times 10^{20} + 0.5 \times 10^{10})B//$

$\qquad\qquad (0.5 \times 10^{20} + 0.5 \times 10^{10})R/(0.5 \times 10^{20} - 0.5 \times 10^{10})B$

and configuration

$(0.5 \times 10^{20} + 0.5 \times 10^{10})R/(0.5 \times 10^{20} - 0.5 \times 10^{10})B//$

$\qquad\qquad (0.5 \times 10^{20} - 0.5 \times 10^{10})R/(0.5 \times 10^{20} + 0.5 \times 10^{10})B.$

It is surely indisputable that no method of analysis would be sufficiently
accurate to distinguish any of these configurations from the most probable
configuration

$(0.5 \times 10^{20})R/(0.5 \times 10^{20})B//(0.5 \times 10^{20})R/(0.5 \times 10^{20})B.$

These calculations establish beyond reasonable doubt that *for values of N
comparable with the number of molecules in a system large enough to be
handled in the laboratory the chance that the assembly be found in any configuration differing perceptibly from the most probable distribution is vanishingly
small.* The same conclusion would have been reached whatever model we
had chosen to study. We assume therefore that it applies to assemblies of
particles of all kinds, so long of course as the number of particles present is
sufficiently great.

One further remark regarding the above results is appropriate. The ratio p/P figuring in our argument may be regarded as a measure of the *mean deviation* from the most probable distribution (in this case as far as ninety five per cent of all complexions are concerned.) The fact that we found experimentally that this mean deviation is of the order $N^{-\frac{1}{2}}$ will come as no surprise to those readers familiar with the mathematics of *fluctuations*. We shall meet similar results in Chapter 8 when we calculate the mean deviation from the mean energy of an assembly in equilibrium with a thermostat, and when in Chapter 12, we calculate the mean deviation from the mean number of particles in an open system in equilibrium with a "reservoir" of particles of the same kind.

The remaining matters to be discussed in this section are purely arithmetical. Table 1.2 gives the total number of complexions and *the number encompassed by the most probable configuration* for the model studied above and for the four values of N considered earlier, and in addition gives the values of the two ratios W_{max}/W_{total} and $\ln W_{max}/\ln W_{total}$.

The point of immediate interest in Table 1.2 is that although for large values of N the value of W_{max} is much less than that of W_{total}, the value for $\ln W_{max}$ is for all practical purposes indistinguishable from that for $\ln W_{total}$. It is a fortunate fact that we shall be concerned only with expressions containing terms in $\ln W_{total}$ (rather than terms in W_{total} itself) so that we can use without significant error $\ln W_{max}$ if this is easier to obtain. As we shall see in later chapters, this does ease some of our studies, in fact in some cases it makes relatively simple, calculations which would otherwise be impossible.

Table 1.2

$N = N_R = N_B$	W_{total}	W_{max}	$\dfrac{W_{max}}{W_{total}}$	$\dfrac{\ln W_{max}}{\ln W_{total}}$
500	2.7×10^{299}	1.35×10^{298}	0.05	0.999
5 000	1.6×10^{3008}	2.5×10^{3006}	0.015	1.000
50 000	1.6×10^{30100}	0.81×10^{30098}	0.003	1.000
500 000	5.6×10^{301026}	1.4×10^{301022}	2.5×10^{-5}	1.000

There is one related aspect of this which must be mentioned. We have already said that for very large values of N we may replace the more accurate form of the Stirling approximation formula

$$\ln N! \approx N \ln N - N + \tfrac{1}{2} \ln (2\pi N) \qquad (1.8)$$

by the simpler expression

$$\ln N! \approx N \ln N - N. \qquad (1.9)$$

If we use expression (1.9) to evaluate $\ln W_{total}$ from equation (1.5) and $\ln W_{max}$ from equation (1.7) we obtain *in each case* the result $2N \ln 2N -$

$2N \ln N$, showing that any error introduced by using $\ln W_{\text{max}}$ instead of $\ln W_{\text{total}}$ is just as insignificant as that introduced by evaluating $\ln N!$ by means of expression (1.9) rather than by means of the mathematically more accurate expression (1.8). (We see now why expression (1.8) was used in all calculations made above: use of (1.9) would have led to the incorrect conclusion that *all* complexions are encompassed by the most probable configuration.)

It is perhaps a fitting conclusion to this section to draw attention to the *magnitude* of the numbers of complexions recorded in the two tables. The *significance* of numbers as great as 10^{301026} or even 10^{299} is quite impossible to comprehend, particularly when it is realised that the number of atoms in our own planet is only about 10^{50}. (Those doubting this figure should study Exercise 1.2.) But numbers as great as 10^{301026} are quite insignificant when compared with the number of complexions accessible to a macroscopic system. For example, the number of complexions accessible to one mole of gas at room temperature and a pressure of one atmosphere is of the order $10^{10^{24}}$, and one of the fascinating aspects of statistical mechanics is that when we use its equations to obtain values for the thermodynamic properties of such systems, although we use machinery which disguises it, we are in effect *counting* the number of complexions concerned.

1.3   Complexions and Quantum-Mechanical States

We must now justify the application of the conclusions reached in the last section to the statistical analysis of a macroscopic chemical system. Since the word *configuration* was used in Section 1.1 when discussing the hydrogen iodide system, and in Section 1.2 when discussing the arrangement of particles on labelled sites in precisely the same sense, *to denote any arrangement of constituent particles which may be identified by macroscopic measurements made on the system as a whole*, so giving the word configuration the same meaning as has the thermodynamic word "state", one point of similarity between the model and the chemical system is immediately apparent, the fact that although very many configurations of both are possible, in each case one configuration (or at least one group of configurations indistinguishable one from the other by macroscopic measurements) is apparently preferred to any other. In the case of the model we found that the reason for this is that the number of *complexions* leading to the most probable configuration greatly exceeds the number leading to any configuration differing perceptibly from it, and it is tempting, and reasonable, to suppose that the reason why, given enough time, any chemical system reaches and remains in one particular configuration is the same, simply the fact that very many more complexions lead to that particular configuration than to any other. In other words we suppose that the equilibrium state reached by any isolated system is indistinguishable from its most probable configuration.

This raises however an essential question. What *are* the different complexions accessible to a chemical system, what are the different arrangements which can be identified only by the examination of each particle in turn? This question is easily answered. Let us suppose that it were possible to subject an isolated system in its equilibrium state to microscopic examination capable of giving information on each and every molecule simultaneously, and that such complete descriptions were repeated many times, say, one a minute for a day. Although macroscopic measurements would show that the thermodynamic properties of the system (its temperature, pressure, composition et cetera) remain the same (because the sytem is in its equilibrium state) these microscopic examinations would reveal continuous changes in its *fine structure*, due to the continuous redistribution of the available energy between one molecule and another as the result of intermolecular collisions, so bringing about continuous changes in the *quantum-mechanical state* of the system as a whole. In other words, because any equilibrium state of a macroscopic system is a state of *dynamic* equilibrium, any thermodynamic state (i.e. configuration) identified by macroscopic measurements carried out on the system as a whole necessarily encompasses many quantum-mechanical states. We suppose therefore that the reason why, given enough time, any isolated macroscopic system eventually attains an equilibrium state is because that particular state encompasses a greater number of quantum-mechanical states than any other.

It is evident however that this hypothesis depends on one assumption. An essential feature of the argument advanced in the study of the model was the supposition that there is no reason why one complexion should be preferred to any other. This appears axiomatic in the case of the model, (just as when rolling a dice the chance of getting a 'one' is the same as that of getting a 'six') but the analogous assumption that in the case of an isolated chemical system all quantum-mechanical states have the same probability most certainly calls for some justification. But in fact no *a priori* justification is available. The formal statement that IN AN ISOLATED SYSTEM, ALL POSSIBLE QUANTUM-MECHANICAL STATES ARE EQUALLY PROBABLE is the basic assumption of statistical mechanics. There is no proof that it is correct. It must be regarded (like the laws of classical thermodynamics) as a postulate which we accept because it enables us to predict results which are confirmed by experiment. In fact this is the philosophy governing the statistical mechanical approach to all problems, and is the immediate answer to any objection that too much weight is placed on an apparent parallel between the behaviour of a well established (exceedingly complex) chemical situation and that involving a relatively simple model. It is good science to put forward reasonable hypotheses in order to develop a theory, and to use the theory to make particular predictions (such as 'theoretical' values of quantities of interest) so long as we *test* the hypotheses by the success or otherwise of the predictions when measured against 'real' experimental results, and reject the hypotheses if they prove to lead to incorrect results.

Emphasis must now be placed on the fact that the basic assumption of statistical mechanics, the equal probability of all accessible quantum-mechanical states holds only if (as stated above) the system is *isolated*, so that its total energy, the volume accessible to its constituent particles, and the amounts of each component are prescribed. We shall not attempt a complete justification of these conditions at present, but the following remarks are pertinent,

 (i) that as we shall see in a later chapter, if the energy of a system is allowed to change all possible complexions are *not* equally probable, the chance of each complexion being reached being *weighted* according to the energy to which it corresponds;

 (ii) that the volume accessible to the system determines the *energy levels* accessible to each of the constituent particles, and therefore affects the number of ways in which the particles may be distributed between them, quite apart from the fact that a change in volume would be accompanied by a change in the energy of the system because of the work term involved; and

 (iii) that the fact that the amounts of all components[1] are prescribed does not prohibit the change of molecules of one species into molecules of another, so that chemical reaction is not ruled out.

We shall denote these conditions by the device (E, V, N prescribed) as in the next paragraph, or, in mathematical expressions by the suffices (E, V, N), N here denoting constancy of the numbers of particles of all components.

We are at last in a position to record a complete statement of the fundamental hypothesis on which statistical mechanics is based, it being that *the equilibrium state reached (given enough time) by any isolated system (E, V, N prescribed) is that which encompasses the greatest possible number of a priori equally probable quantum-mechanical states.*

Before proceeding it is perhaps as well to ensure that the meaning of some of the most important terms used so far in our presentation is entirely clear.

[1] We are of course using the word *component* in the classical sense, meaning the minimum number of substances which must be available in order to assemble any chosen mixture of the molecular species in question. For a very complete discussion of the relationship between the number of components and the number of distinct molecular species in a chemical system, the reader is referred to Denbigh, *Principles of Chemical Equilibrium* (Cambridge University Press 1957, pp. 167–170).

First the word *assembly*, which we have used to denote the collection of particles with which, in any problem, we are concerned, and which therefore has precisely the same meaning as has the word *system* as used in thermodynamics, and the two terms may be used indiscriminately,[1] although we shall tend to use the word assembly more frequently simply because it carries with it the sense of a collection of individual particles which is the way statistical mechanics views any system. Next the word *configuration*, which means any arrangement of the constituent particles which may be identified by macroscopic measurements made on the system (assembly) as a whole, and which therefore has (as has been emphasised more than once in earlier pages) precisely the same meaning as has the word *state* as used in classical thermodynamics. Third, the word *complexion* meaning each of those detailed descriptions of the assembly made possible only by examination of each particle in turn, the number of complexions associated with a physical system being the number of quantum-mechanical states accessible to it. The last term, *quantum-mechanical states* is slightly more difficult because it is foreign to classical thermodynamics and arises in quantum mechanics. It is perhaps opportune to say here, that although quantum mechanics is prerequisite to statistical mechanics a reasonably satisfactory introductory account can be presented requiring only acceptance of the principle of the quantisation of energy and that of the quantum-mechanical concepts of the *energy levels* accessible to any particle in the assembly, the *degeneracy* of each energy level (this being the number of distinct quantum states associated with each energy level), and the quantum-mechanical states accessible to the assembly as a whole. The meaning of all these terms will become clear when, in later chapters we study definite physical systems. In the meantime it is probably best not to attempt to improve on the meaning of the term complexion already implied, *each of those detailed descriptions of the assembly resulting from the continuous redistribution of the available energy between one particle and another.*

1.4 The Criterion for Spontaneous Change, the Approach to the Second Law

One consequence of the fundamental hypothesis deduced in the last section is immediately apparent. All configurations (thermodynamic states) accessible to an isolated system other than the most probable configuration (the equilibrium state) must encompass *less* than the number of complexions

---

[1] Most authors use the word assembly in statistical mechanics to mean the collection of particles concerned, but some use the word system to denote each of the particles composing the assembly, i.e. they speak of "an assembly of systems." We prefer to restrict the word "system" to that meaning given it in thermodynamics.

(quantum-mechanical states) encompassed by the most probable configuration, and *since all spontaneous processes are towards equilibrium, any such process in an isolated system must be accompanied by an increase in the number of possible complexions (quantum-mechanical states).*[1] This statement obviously approaches very closely indeed the province of the second law of thermodynamics, but before pursuing this line let us see whether the statement is borne out by the study of a simple physical process, the adiabatic expansion of a gas.

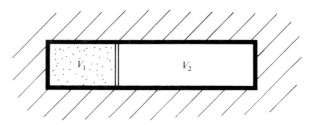

Figure 1.3

Consider a closed vessel divided by a barrier, as shown in Figure 1.3, into two parts of volumes V_1 and V_2. Suppose that initially N molecules of gas are confined to volume V_1, that V_2 is empty and that the system is thermally insulated from the surroundings. The system (all that enclosed by the heavy lines in the figure) is completely isolated from the surroundings so that in any subsequent action the total energy, the total volume ($V_1 + V_2$) and the number of molecules of gas are conserved. Suppose now that the barrier is removed so that the gas is free to diffuse into the total volume. What is the chance that all molecules will remain in the original part of the vessel V_1 (configuration 1) rather than that they will spread over the total volume $V_1 + V_2$ (configuration 2)?

It seems reasonable to suppose that the chance that at some instant in time *one* particular molecule is located in the smaller volume is $V_1/(V_1 + V_2)$, and that therefore the chance that all molecules remain in V_1 rather than diffuse over the whole volume available is $[V_1/(V_1 + V_2)]^N$. Denoting the chance of finding the assembly in configuration 1 by the symbol ϖ_1 and that of finding it in configuration 2 by the symbol ϖ_2 we see that

$$\frac{\varpi_2}{\varpi_1} = \left(\frac{V_1 + V_2}{V_1}\right)^N$$

[1] We have in this statement placed together the terms 'configurations' and 'thermodynamic states', the terms 'the most probable configuration' and 'the equilibrium state' and the terms 'complexions' and 'quantum-mechanical states' so as to emphasise the points made in the last paragraph of the last section, but for the remainder of this chapter we shall use only the terms 'configuration', 'the most probable configuration' and 'complexion'.

showing that ϖ_2 is greater than ϖ_1 which of course, we expect. Earlier we deduced that the chance that a particular configuration is reached by an isolated system is proportional to the number of complexions encompassed by it, so that denoting the number of complexions encompassed by configuration 1 by the symbol W_1 and the number encompassed by configuration 2 by the symbol W_2 we see that

$$\frac{W_2}{W_1} = \frac{\varpi_2}{\varpi_1} = \left(\frac{V_1 + V_2}{V_1}\right)^N$$

so reaching the conclusion that W_2 is necessarily greater than W_1. That this conclusion is more than reasonable is seen immediately from the fact that W_2 must necessarily *contain* W_1, because even when the larger volume is accessible, all molecules *may* if so inclined, remain in the smaller volume V_1 (so spreading themselves over only W_1 complexions) and their passage into the additional volume V_2 cannot reduce the number of possible complexions but can either open up more or leave the number unchanged. Two other points should be mentioned. The first is that as we shall see in a later chapter, quantum mechanics shows that increasing the volume accessible to an assembly of non-localised particles (a gas is such an assembly) increases the number of energy levels accessible to each particle, and so necessarily increases the number of ways in which a fixed number of particles may be distributed over accessible energy levels. The second is more general. The barrier originally confining the molecules to volume V_1 may be thought of as a *restraint* preventing them from occupying the larger volume. It is a general result in statistical mechanics that the lifting of any restraint leads to an increase in the total number of possible complexions.

The above discussion shows that it may reasonably be inferred that *a spontaneous change in an isolated system is accompanied by an increase in the number of accessible complexions, and that the configuration eventually chosen by the system is that encompassing the maximum number of complexions.* Classical thermodynamics however teaches that *a spontaneous change in an isolated system leads to an increase in its entropy and that the equilibrium state eventually attained is that at which its entropy is a maximum.* These statements suggest that the W value[1] and the entropy of a system are closely related. Let us suppose that one is a function of the other and write

$$S = f(W).$$

We must now deduce the type of function involved.

Consider two independent sub-systems A and B situated side by side, the number of complexions of the first being W_A and that of the second W_B.

---

[1] When for convenience, we refer to the W value of a system we mean the *total* number of possible complexions, i.e. the sum of the W values of all accessible configurations.

What is the number of complexions accessible to the two sub-systems viewed as *one* system? To each of the complexions accessible to A there are W_B complexions accessible to B, so that the number accessible to the whole is $W_A \times W_B$. Entropy is however an extensive property, and that of the combined system is the sum of the entropies of its parts. This suggests the logarithmic relationships[1]

$$S_A = k \ln W_A, \qquad S_B = k \ln W_B$$

and

$$S_{A,B} = S_A + S_B = k \ln (W_A \times W_B) = k \ln W_{A,B},$$

where k is some constant the value of which is the same for each part and for the whole. We thus infer that the entropy of a system is related to the number of complexions accessible to it by the equation

$$S = k \ln W. \tag{1.10}$$

This, known as the Boltzmann equation, may be regarded as the *bridge* connecting statistical mechanics and classical thermodynamics. Although final acceptance of its reality must await more quantitative investigation, we will assume its soundness and draw attention to some of its consequences.

Let us suppose that in a particular case one complexion only is accessible to the system, so that its W value is unity. Our knowledge of the system's fine structure is then complete. If on the other hand the W value is two, our knowledge of its fine structure is somewhat uncertain, it may be in one complexion or it may be in the other. As the W value of the system increases, our detailed knowledge of its fine structure decreases, all we can do is to surmise the chance that it is in any particular complexion. We describe the situation when only one complexion is accessible, the situation in which our knowledge of its fine structure is complete as a state of complete order. When more than one complexion is accessible a state of less than complete order exists. This argument is of course the basis of the familiar

---

[1] It may appear that these relations are unnecessarily restrictive, since the W values and entropies would also be satisfied by the more general relations

$$S_A = k \ln W_A + x_A, \qquad S_B = k \ln W_B + x_B$$

and

$$S_{A,B} = k \ln (W_A \times W_B) + x_A + x_B,$$

where x_A and x_B are arbitrary constants. We prefer to ignore this possible complication at present, as it can be considered more profitably in Chapter 5. In the mean-time it should be accepted that ignoring these arbitrary constants (in effect assigning them the value zero) is in accord with that version of the third law of thermodynamics which states that the entropies of all pure crystalline bodies may be assigned the value zero at the absolute zero of temperature.

statement that an increase in the entropy of an isolated system is equivalent to an increase in its disorder.

1.5   The Statistical–Mechanical Statement of the Second Law

In classical thermodynamics the most elegant and vivid expression of the second law is the statement that for a closed, isolated system

$$\left(\frac{\partial S}{\partial \xi}\right)_{E,V,n} \geq 0 \tag{1.11}$$

where ξ is the extent of any possible process, so that $\partial \xi$ denotes an infinitesimal progress of the process, the suffices E, V, n express constancy of the energy and volume of the system and the amounts of its constituent components, and where the inequality sign governs the feasibility of the process and the equality sign provides the condition for equilibrium. Equation (1.10) leads immediately to the statistical analogue of expression (1.11)

$$\left(\frac{\partial \ln W}{\partial \xi}\right)_{E,V,N} \geq 0 \tag{1.12}$$

where all symbols carry the same significance as before. (We have however changed the suffix n in (1.11) where the amounts of material are usually expressed in moles, to N denoting as always, the number of particles of each component.)

Just as (1.11) can be extended to cover the interaction of two systems A and B, (as long as the pair are isolated from the rest of the universe) to give the expression

$$\left(\frac{\partial (S_A + S_B)}{\partial \xi}\right)_{E(=E_A+E_B),\,V(=V_A+V_B),\,n(=n_A+n_B)} \geq 0 \tag{1.13}$$

(showing incidentally that when two bodies interact spontaneously the entropies of both may increase, but also, and in fact more usually, that the entropy of one may decrease as long as this is more than compensated by the corresponding increase in the entropy of the other), so may (1.12) be extended to give the expression

$$\left(\frac{\partial \ln (W_A \times W_B)}{\partial \xi}\right)_{E(=E_A+E_B),\,V(=V_A+V_B),\,N(=N_A+N_B)} \geq 0 \tag{1.14}$$

(showing that the W values of both bodies may increase, but that one may decrease as long as this is more than compensated by the corresponding increase in the W value of the second), and these expressions may of course, be further extended to cover the interaction of a group encompassing any number of bodies, as long as the group as a whole is isolated from the surroundings. Indeed, since the universe (which by definition includes *all* bodies) is in effect isolated, just as (1.11) can be extended to give the attractive (if not particularly useful) statement that all events lead to an increase in the entropy of the universe (made originally by Clausius, "Die Energie der Welt ist konstant; die Entropie der Welt strebt einem Maximum zu") so may (1.12) be extended to give the equally attractive (if not particularly useful) statement that all events lead to an increase in the W value of the universe.

The statistical foundation of the second law can however be completed only by drawing attention to one rather subtle point of difference between the thermodynamic and statistical-mechanical views of it, which can perhaps best be explained by considering the possibility of an isolated system consisting of a perfect mixture of two gases A and B, (say N molecules of each) undergoing spontaneously a process resulting in complete separation, a process shown diagrammatically in Figure 1.4. Classical thermodynamics says that spontaneous separation *cannot* happen because it would lead to a decrease in entropy equal to $(2N/L)R \ln 2$. Statistical mechanics says that it *may* happen, but that the chance of it happening is so small (actually[1] $(\frac{1}{2})^{2N}$) that the event is extremely *unlikely*.[2] There is one point on which we should be clear. The spontaneous separation would *not*, if it occurred lead to a decrease in W, and so conflict with expression (1.12). It would in fact lead to no change in W, as the total number of available complexions would remain the same. The spontaneous separation would be a manifestation of the fact that the system had chanced to *relinquish* the overwhelmingly large proportion of the total number of complexions accessible to it, not that that proportion has become inaccessible. The W value could be reduced only as the result of action by an outside agent inserting a barrier between the two

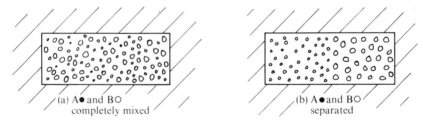

(a) A● and B○ completely mixed   (b) A● and B○ separated

Figure 1.4

[1] The reader should satisfy himself regarding the correctness of these values.

[2] Planck, in an essay entitled "My Scientific Autobiography" tells of the extreme antagonism which developed between Clausius (the inventor and exponent of the entropy principle) and Boltzmann (the founder of statistical mechanics) as the result of this difference in point of view.

species when in the state shown in (b), (so preventing the species mixing again) and so making a great proportion of the hitherto accessible complexions inaccessible.

1.6   The Identification of k (the Boltzmann Constant)

It was established earlier that the quantity k appearing in equation (1.10) is some constant the value of which is the same for all systems. We can therefore establish its value by studying the behaviour of any system chosen only for convenience. Consider the process discussed in Section 1.4, the free expansion of N molecules (N/L moles, L being the Avogadro constant) of perfect gas from volume V_1 to ($V_1 + V_2$). We earlier showed that the W values of the initial and final states are related by the equation

$$\frac{W_2}{W_1} = \left(\frac{V_1 + V_2}{V_1}\right)^N.$$

It follows from (1.10) that the change in entropy of the system is

$$\Delta S = k \ln W_2 - k \ln W_1 = kN \ln \left(\frac{V_1 + V_2}{V_1}\right).$$

But classical thermodynamics shows that such a process leads to an entropy change given by the equation

$$\Delta S = \frac{N}{L} R \ln \left(\frac{V_1 + V_2}{V_1}\right),$$

where R is the molar gas constant. It follows that

$$k = R/L \tag{1.15}$$

The constant k is known as the Boltzmann constant, and its value is 1.3804×10^{-23} JK^{-1}.

1.7   Other Consequences of the Equation $S = k \ln W$

We must not permit the facility with which this equation was reached to disguise the far-reaching effect of the step, because this relationship gives not only a statistical interpretation of entropy but leads also to statistical expressions for other thermodynamic functions.

We have assumed that the W value for a system depends on its energy, its volume and the number of particles of each component, which assumption can conveniently be denoted by the expression

$$W = f(E, V, N_1, N_2 \ldots) \tag{1.16}$$

where f is some function the form of which is not necessarily known. It follows from (1.16) that

$$
\begin{aligned}
d \ln W = {} & \left(\frac{\partial \ln W}{\partial E}\right)_{V,N_1,N_2} dE + \left(\frac{\partial \ln W}{\partial V}\right)_{E,N_1,N_2} dV \\
& + \left(\frac{\partial \ln W}{\partial N_1}\right)_{E,V,N_2} dN_1 + \left(\frac{\partial \ln W}{\partial N_2}\right)_{E,V,N_1} dN_2 + \cdots
\end{aligned}
\tag{1.17}
$$

But from classical thermodynamics

$$dS = \frac{dE}{T} + \frac{P}{T} dV - \frac{\mu_1}{T} dN_1 - \frac{\mu_2}{T} dN_2 - \cdots \tag{1.18}$$

where μ_1 and μ_2 are the chemical potentials[1] of components 1 and 2. Comparison of (1.17) and (1.18) shows that since we have identified S with $k \ln W$, i.e. dS with $kd \ln W$, it automatically follows that

$$\left(\frac{\partial \ln W}{\partial E}\right)_{V,N_1,N_2} = \frac{1}{kT} \tag{1.19}$$

$$\left(\frac{\partial \ln W}{\partial V}\right)_{E,N_1,N_2} = \frac{P}{kT} \tag{1.20}$$

$$\left(\frac{\partial \ln W}{\partial N_1}\right)_{E,V,N_2} = -\frac{\mu_1}{kT} \tag{1.21}$$

and so on. These relationships, correctly to be regarded as *equivalent* to equation (1.10) (equivalent in the sense that if (1.10) is true the others are true), are so important that they call for immediate investigation.

Consider the following cases;

(a) *A process concerned with the attainment of thermal equilibrium*

Suppose that a sub-system A containing N molecules in a fixed volume V_A and possessing energy E_A, and a sub-system B containing M molecules in a

---

[1] The terms μ_1 and μ_2 in (1.18) are defined by the equations

$$\mu_1 = (\partial G/\partial N_1)_{T,P,N_2,\ldots} \quad \text{and} \quad \mu_2 = (\partial G/\partial N_2)_{T,P,N_1,\ldots}$$

and are the partial *molecular* free energies, whereas in thermodynamics it is more usual to use partial *molar* free energies defined by the equation $\mu_i^* = (\partial G/\partial n_i)_{T,P,n_j}$ where n_i is the amount of i expressed in moles. Partial molecular quantities are always used in statistical mechanics, and are of course related to the classical quantities by the equation $L\mu_i = \mu_i^*$. Both quantities are correctly called the chemical potential of i.

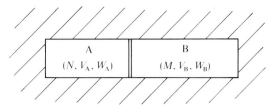

Figure 1.5

fixed volume V_B and possessing energy E_B are brought into thermal contact and then isolated from the surroundings as shown in Figure 1.5. Since we have *not* stipulated that the initial temperatures of the sub-systems are the same, the only possible process that can occur is the passage of heat from one to the other, changing the energies (and W values) of each. It follows from (1.14) that the state of thermal equilibrium eventually reached must be governed by the condition

$$\left(\frac{\partial \ln W_A \times W_B}{\partial E_A} \right)_{E,V_A,V_B,N,M} = 0 \tag{1.22}$$

where for an infinitesimal process dE_A is the increase in energy of A. Since the total energy E is fixed, $dE_A = -dE_B$, so that (1.22) may be written in the form

$$\left[\frac{\partial \ln W_A}{\partial E_A} - \frac{\partial \ln W_B}{\partial E_B} \right]_{E,V_A,V_B,N,M} = 0 \tag{1.23}$$

which expression must be the statistical criterion for thermal equilibrium between A and B. But from the classical zeroth law, the criterion for thermal equilibrium is equality of temperature. The left-hand term must therefore represent a statistical measure of the temperature of A, and the right-hand term a statistical measure of the temperature of B, so that for any system the expression $(\partial \ln W/\partial E)_{V,N}$ must represent a statistical measure of temperature, which is just what is implied by equation (1.19).

(b) *A process concerned with the attainment of mechanical equilibrium*

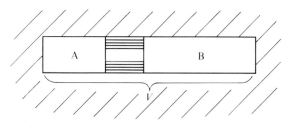

Figure 1.6

Consider two sub-systems A and B consisting respectively of N molecules of a gas of one sort and M molecules of a gas of another sort contained in a closed cylinder of fixed volume V fitted with a frictionless diathermic piston initially locked in some position as shown in Figure 1.6, and suppose that sub-system A is situated entirely in the volume to the left of the piston and sub-system B entirely in the volume to the right. Suppose that the system is isolated, and that thermal equilibrium has been established between A and B.

Suppose that the piston is now unlocked. Since we have *not* stipulated that the sub-systems are initially at the same pressure, one sub-system may expand at the expense of the other. We will suppose that mechanical equilibrium is established when the piston reaches a position so that the volume of sub-system A is V_A and that of sub-system B is V_B. It follows from (1.14) that the final equilibrium state is governed by the condition

$$\left(\frac{\partial \ln W_A \times W_B}{\partial V_A}\right)_{E,V=V_A+V_B,N,M} = 0 \tag{1.24}$$

where dV_A is the increase in the volume of A resultant on an infinitesimal movement of the piston. Since the total volume is constant, $dV_A = -dV_B$, so that (1.24) can be written in the form

$$\left[\frac{\partial \ln W_A}{\partial V_A} - \frac{\partial \ln W_B}{\partial V_B}\right]_{E,V,N,M} = 0 \tag{1.25}$$

Equation (1.25) must therefore be the statistical criterion for mechanical equilibrium between A and B. But the classical criterion is equality of pressure. The left-hand term in (1.25) must therefore represent a statistical measure of the pressure of sub-system A, and that on the right a measure of the pressure of sub-system B which accords with equation (1.20).

(c) *A process concerned with the attainment of phase equilibrium*

Consider now the attainment of equilibrium when a solid sublimes in a closed isolated vessel. Suppose that a solid containing N molecules is introduced into the vessel and that M molecules pass into the gas phase. It follows from (1.14) that the equilibrium finally established is governed by the equation

$$\left(\frac{\partial \ln W_s \times W_g}{\partial M}\right)_{E,V,N} = 0 \tag{1.26}$$

where W_s is the number of complexions corresponding to the solid phase and W_g the number corresponding to the gas. Since the number of molecules remaining in the solid is $N - M$ and since $dM = -d(N - M)$, the condition governing the final equilibrium state may be written in the form

$$\left[-\frac{\partial \ln W_s}{\partial(N-M)} + \frac{\partial \ln W_g}{\partial M} \right]_{E,V,N} = 0 \tag{1.27}$$

i.e.

$$\left[\frac{\partial \ln W_s}{\partial N_s} - \frac{\partial \ln W_g}{\partial N_g} \right]_{E,V,N} = 0 \tag{1.28}$$

where we have denoted the number of molecules in the solid phase by the symbol N_s and the number in the gas phase by the symbol N_g. The classical condition for interphase equilibrium is simply equality of the chemical potentials of each species present in each phase. The left-hand term in (1.28) must therefore represent a statistical measure of the chemical potential of the species in the solid phase, and the term on the right a statistical measure of the chemical potential of the species in the gas phase, which is just what is implied by equation (1.21).

1.8   The Conceptual Foundations Revisited

The object here is to draw together the most important conclusions reached in the preceding sections.

The point on which our treatment of statistical mechanics depends is the experimental fact that *any* closed macroscopic system maintained under prescribed conditions for long enough reaches a particular time-independent *state* or *configuration* (which may be identified by macroscopic measurements giving its temperature, pressure, composition and so on) in which it appears to remain as long as the conditions are maintained. This apparently time-independent configuration is known as *the equilibrium state.* We chose to consider the equilibrium states reached by systems which are isolated from the rest of the universe, and we examined in detail the equilibrium mixture reached as the result of the partial decomposition of hydrogen iodide. We might have examined other systems. One of particular simplicity and relevance to chemistry would reveal the fact that if N gas molecules are placed in a container divided into two parts of volume V_1 and V_2 respectively connected by a small hole, the system chooses a configuration in which N^* molecules occupy V_1 and $(N - N^*)$ occupy V_2, N^* being given by the equation[1]

$$\frac{N^*}{N - N^*} = \frac{V_1}{V_2}.$$

---

[1] The statistical–mechanical foundation of this equation is the subject of Exercise 9.1.

Another of equal relevance is the fact that if a liquid is placed in a closed isolated container, a particular number of molecules (depending on the volume available and the energy of the system) choose to enter the gas phase. We would obtain a similar result whatever system we studied.

Whereas classical thermodynamics exhibits no curiosity as to *why* any system moves towards an equilibrium state, but is content to identify it by applying the appropriate thermodynamic criterion, statistical mechanics seeks the reason for it, and advances the theses:

(i) that this behaviour is consequent on the fact that all macroscopic physical systems are assemblies of very large numbers of dynamic particles, so that the system is continuously passing from one configuration to another,

(ii) that each possible configuration encompasses a particular number of detailed arrangements or *complexions*, the number of complexions being identified (somewhat tentatively at this stage) with the number of possible distinguishable quantum-mechanical states,

(iii) that for a closed system of prescribed energy and volume, each possible complexion has the same statistical weight, (which simply means that one complexion is just as likely to arise as is any other.)

Assumption (iii) means that the *chance* that a system reaches a particular configuration is proportional to the number of complexions encompassed by it, so that if the number of complexions (denoted by the symbol W_{max}) leading to a particular configuration Y exceeds that leading to any other configuration, configuration Y may justifiably be called the *most probable configuration*.

Studies of any model containing a very large number of potentially mobile particles show indeed that the number of complexions leading to the most probable configuration so greatly exceeds the number leading to any configuration differing perceptibly from it, that the chance that macroscopic analysis would reveal the assembly in any but the most probable configuration is vanishingly small. In other words, statistical mechanics identifies the equilibrium state as being (or at least indistinguishable from) the most probable configuration.

We take it therefore that if as the result of the lifting of some restraint an isolated system changes its state, the process is one leading to an increase in the number of complexions accessible to the system, so that any spontaneous process at fixed E, V, N is governed by the expression

$$\left(\frac{\partial W}{\partial \xi}\right)_{E,V,N} > 0$$

where W is the number of complexions and ξ the extent of reaction, and that the equilibrium state finally reached is identified by the equation

$$\left(\frac{\partial W}{\partial \xi}\right)_{E,V,N} = 0.$$

(The fact that in practice we prefer to consider the change in ln W rather than the change in W itself is of course immaterial.)

These expressions, which we take to be the statistical–mechanical foundation of the second law of thermodynamics, permit a statistical expression for the entropy of a system, the equation

$$S = k \ln W,$$

where k is a universal constant shown to equal R/L. This equation should be regarded as the *bridge* between classical thermodynamics and statistical mechanics.

The argument so far has been tentative, and all that has been shown so far (by the analyses carried out in the last section) is that it is internally self-consistent, and that it leads to reasonable statistical interpretations of such phenomena as the establishment of thermal, mechanical and phase equilibria, but the theory must stand or fall on the degree of success achieved when it is used to make quantitative predictions which may be compared with real experimental values. The theory will be so tested, first by being applied to an investigation of the properties of an atomic crystal in Chapters 2 and 3, and then to a study of the properties of a perfect gas in Chapters 4 and 5.

EXERCISES

1.1   Using the fact that ΔH for the reaction $H_2 + I_2 \rightarrow 2HI$ at 300 K is -10.38 kJ, and making use of two assumptions (both in fact very nearly true) one that the value of ΔH is independent of temperature, and the other that the heat capacity at constant volume of a mixture of one mole of hydrogen and one mole of iodine vapour is 42 JK^{-1} and is independent both of the temperature and of the extent of reaction, show that the various configurations described on page 5 are indeed states of equal energy.

1.2   The mass of the earth is about 6×10^{24} kg. The number of electrons, protons and neutrons on earth are about the same, the mass of a proton (or neutron) is 1.6×10^{-27} kg and that of an electron negligible. The distribution of the various elements throughout the earth appears to be such that the "mean atomic weight" is about 20, corresponding to ten neutrons, ten protons and ten electrons. Show (i) that the number of fundamental particles on earth is about 5×10^{51}, and (ii) that the number of atoms on earth is of the order 10^{50} as stated on page 18.

1.3   Calculate ln $N!$ for $N = 10$, first by evaluating $N!$ by direct multiplication and second evaluating ln $N!$ direct by using the Stirling formula (1.8), and compare the results. Repeat for $N = 20$ to satisfy yourself of the negligibility of any error resulting from the use of (1.8) for large numbers.

1.4 Calculate $\ln N!$ for $N = 10$ first using expression (1.8) and secondly using (1.9). Repeat for larger numbers to satisfy yourself that any error introduced by using (1.9) rather than (1.8) is negligible for numbers as large as 10^{20}.

1.5 Consider the model described in Section 1.2 for $N = N_R = N_B = 500$, 5000, 50 000, and 500 000. Using (1.8) to evaluate the factorials, (i) calculate the total number of complexions in each case (from equation 1.5), and (ii) set up a computer program which evaluates W_n (from equation 1.6) for values of $n = N/2, N/2 \pm 1, N/2 \pm 2, \ldots$, and so on, and show that ninety per cent of all complexions are encompassed by only 27, 83, 261 and 817 configurations centred around the most probable configuration. Express these figures as a percentage of the total number of configurations (consult the footnote on page 13) in each case.

CHAPTER 2

An Assembly of Independent Localised Particles

We now need to put the ideas discussed in the last chapter onto a firmer basis, so that they may be used to throw light on the behaviour of real systems. The procedure adopted is as follows: rather than attempt to apply the ideas to a real system directly, we apply them to an idealised analogue of it, a *model*, the principal features of which appear to correspond reasonably closely to those of the system of interest, but which is simple enough for its analysis to be amenable to the methods at our disposal. We thus obtain expressions leading to theoretical values for the properties of the model for comparison with the corresponding experimental values for those of the real system. The method is of course open to two sources of error. Some feature of the model may differ from the corresponding feature of the real system in some way which is more important than we first suppose, and even if this is not so its analysis will almost certainly entail the use of mathematical approximations some of which may prove to be unjustified. The model and its mathematical analysis provide however, their own justification. If no correspondence is obtained between the theoretical values for the properties of the model and the experimental values for those of the real system, the model (or its mathematical treatment) must of course be abandoned. If the theoretical and experimental results show the same general trends but lack precise quantitative agreement it may be possible to determine which aspects of the model or which mathematical approximations used in its analysis are unsound, and so point the way to a modification leading to closer correspondence. Where however, close agreement between theoretical and experimental values is obtained, we can be reasonably confident that the correspondence between model and system is equally close, and since we presumably 'understand' the behaviour of the model, we achieve, in some degree, an understanding of the behaviour of the real system.

The aim of this chapter and of the next is to obtain insight into the behaviour of an atomic crystal. In this chapter we devise the model set up to represent it, and obtain expressions for the thermodynamic properties it would exhibit. In Chapter 3 we describe the attempts which have been made to evaluate these expressions when applied to the real system.

2.1　The Model

As just stated, our first task is to devise a model which corresponds reasonably closely to an atomic crystal. That which we choose is an assembly of N identical particles (we shall in this chapter always use the word particle rather than either of the words atom or molecule to emphasise that we are considering a model rather than a real system) which, we suppose, are *localised* in space, and *independent* but able to exchange energy. Both aspects of these descriptions require further discussion. By supposing that the particles are localised, we imply that each is confined to a small region of space (a cell bounded by its neighbours) but free to move within that space. We may therefore suppose that each particle is 'tethered' so that it can oscillate about its mean position but is incapable of diffusing from one cell to another. (This is of course equivalent to supposing that the particle possesses vibrational but not translational energy.) The fact that each particle is supposed to occupy a definite position in the assembly means that although all are identical, each is so to speak 'labelled' by its position and so *distinguished* (in principle at least) from any other like particle occupying a different position. The model consists therefore of an assembly of *identical* but *distinguishable* particles.

We now turn to the word 'independent'. If we were to attempt to study any assembly subject to strong inter-particle forces, not only would it be impossible to conceive the behaviour of any one particle without interference from its neighbours, but the expression for the total energy of the assembly would include potential energy terms relating to the inter-particle forces, and the resulting complications would be so great that they would put the exercise without the scope of the present treatment. Instead, we suppose that although energy can be passed from one particle to another the *state* of any one is at any moment, entirely independent of the state of all other particles in the assembly, so that the behaviour of any one can be studied without reference to the remainder, and so that the total energy of the assembly can be expressed as the sum of the 'private' energies of each particle therein.

In accordance with quantum theory we suppose that only discrete energy levels $\epsilon_0, \epsilon_1, \ldots, \epsilon_i, \ldots$ are accessible to each particle, where the lowest possible level is designated ϵ_0, the next $\epsilon_1, \ldots$ and so on, but the fact that we suppose that energy can pass from one particle to another presumably means that the number of particles in each energy level may continually change, there being at one moment

$$n_0 \text{ particles at level } \epsilon_0,$$
$$n_1 \text{ at level } \epsilon_1,$$
$$\vdots$$
$$n_i \text{ at level } \epsilon_i,$$
$$\vdots$$

but that at another moment there are

$$n_0' \text{ particles at level } \epsilon_0,$$
$$n_1' \text{ at level } \epsilon_1,$$
$$\vdots$$
$$n_i' \text{ at level } \epsilon_i,$$
$$\vdots$$

at another

$$n_0'' \text{ at level } \epsilon_0,$$
$$n_1'' \text{ at level } \epsilon_1,$$
$$\vdots$$
$$n_i'' \text{ at level } \epsilon_i,$$
$$\vdots$$

and so on. If however we impose the condition that the assembly is *closed* and *isolated* as we must do if we are to use the ideas put forward in the last chapter, each 'set' of population numbers must be governed by two conditions, one,

$$E = n_0\epsilon_0 + n_1\epsilon_1 + \cdots + n_i\epsilon_i \cdots$$
$$= n_0'\epsilon_0 + n_1'\epsilon_1 + \cdots + n_i'\epsilon_i \cdots$$
$$= \cdots$$

i.e.

$$E = \sum_i n_i\epsilon_i = \sum_i n_i'\epsilon_i = \cdots \tag{2.1}$$

and the other

$$N = n_0 + n_1 + \cdots + n_i \cdots$$
$$= n_0' + n_1' + \cdots + n_i' \cdots$$
$$= \cdots$$

i.e.

$$N = \sum_i n_i = \sum_i n'_i = \cdots .\qquad(2.2)$$

Each 'set' of population numbers obviously results in a different configuration for the assembly, but since each particle is distinguished by its position, each configuration may be reached through many different detailed distributions (complexions) depending on which particles are at each energy level. It therefore seems reasonable to suppose that the total number of complexions accessible to the assembly is the sum of the number of ways in which each configuration may be reached, that is to say, the number of ways in which the total energy can be distributed so that there are n_0 particles at energy level ϵ_0, n_1 at level $\epsilon_1, \ldots, n_i$ at level $\epsilon_i, \ldots$ plus the number of ways in which the energy can be distributed so that there are n'_0 particles at energy level ϵ_0, n'_1 at level $\epsilon_1, \ldots, n'_i$ at level $\epsilon_i, \ldots$ and so on, all distributions being governed by conditions (2.1) and (2.2).

2.2

Before attempting what at first sight appears to be an almost impossible problem, let us consider a much simpler exercise, the determination of the number of complexions resulting from the distribution of four units of energy among five particles.

It is immediately apparent that if the particles are supposedly indistinguishable, the energy can be distributed in only five ways:

  (I) one particle having four units and four none,

 (II) one having three units, one one and three none,

(III) two having two units and three none,

(IV) one having two units, two one and two none, and

 (V) four having one unit and one none.

But if the particles are distinguished one from another (i.e. supposedly labelled A, B, C, D and E), instead of there being only one identifiable complexion corresponding to (I) there will be five,

$$A(4), \quad B, C, D, E(0)$$
$$B(4), \quad A, C, D, E(0)$$
$$C(4), \quad A, B, D, E(0)$$
$$D(4), \quad A, B, C, E(0)$$
$$E(4), \quad A, B, C, D(0),$$

and instead of there being only one identifiable complexion corresponding to (II) there will be twenty,

$$A(3), \quad B(1), \quad C, D, E(0)$$
$$A(3), \quad C(1), \quad B, D, E(0) \quad \text{and so on.}$$

In fact, instead of there being only five distinguishable complexions (I) to (V), there are seventy in all, five leading to (I), twenty to (II), ten to (III), thirty to (IV) and five to (V). Let us see how these figures arise using the principles of combinatorial mathematics. We can think of any one particle as falling into any one of five possible 'groups' according to the number of units of energy it acquires, the groups being designated (0), (1), (2), (3) and (4). In (I) we have four particles in group (0), none in groups (1), (2) and (3), and one in group (4). The number of ways in which five such particles can be divided into five such groups is

$$\frac{5!}{4!\,0!\,0!\,0!\,1!}$$

which, remembering that 0! is unity gives five. In (II) we have three particles in group (0), one in group (1), none in group (2), one in group (3) and none in group (4). The number of ways in which five such particles can be divided into five such groups is

$$\frac{5!}{3!\,1!\,0!\,1!\,0!}, \quad \text{i.e. twenty.}$$

The same procedure leads to our finding ten complexions corresponding to (III), thirty to (IV) and five to (V). The total number of complexions encompassed by all possible distributions is therefore given by the expression

$$W_{\text{total}} = W_{(\text{I})} + W_{(\text{II})} + W_{(\text{III})} + W_{(\text{IV})} + W_{(\text{V})}$$

$$= \sum_{(\text{I})}^{(V)} \frac{5!}{\prod_i n_i!}$$

where $\prod_i$ denotes the running product of all possible values of $n_i!$ in each configuration, each configuration being of course subject to the condition that the total number of particles and the total number of units of energy available for distribution are conserved.

2.3

Returning to the main exercise it now appears that the total number of complexions accessible to an assembly of N localised (i.e. distinguishable) particles (the energy being fixed) is given by the expression

$$W_{total} = \frac{N!}{\Pi_i n_i!} + \frac{N!}{\Pi_i n_i'!} + \cdots \tag{2.3}$$

subject to the conditions

$$E = \sum_i n_i \epsilon_i = \sum_i n_i' \epsilon_i = \cdots \tag{2.1}$$

and

$$N = \sum_i n_i = \sum_i n_i' = \cdots . \tag{2.2}$$

Before proceeding, one correction to equation (2.3) must be made. This equation clearly depends on the assumption that the state of each particle is determined solely by the energy it possesses, in quantum-mechanical language, that all energy levels are non-degenerate. This assumption is incorrect. Almost all energy levels accessible to real particles are degenerate (i.e. more than one quantum state of the particle is characterised by the same energy) and this has a considerable bearing on the equations we shall derive. We could have introduced the question of degeneracy at the start of our argument, but it is simpler conceptually to proceed as we have done and to introduce the necessary correction terms at this stage.

The concept of degeneracy emerges in quantum mechanics, being the number of different solutions associated with the same energy level to the Schrödinger wave equation, each solution corresponding to a different quantum state, but all that we require at present is contained in the statement that if g different quantum states accessible to a particle are characterised by the same energy ϵ_i, that energy level is said to be a g-fold degenerate. We must now show how to amend equation (2.3) so as to take possible degeneracies into account.

If each of the n_i particles occupying energy level ϵ_i can be found in g_i distinguishable states, then, since the number of ways in which n_i particles can be allotted among g_i states is $g_i^{n_i}$, the number of complexions resulting from the distribution of energy E among N particles so that there are n_0 in energy level ϵ_0, n_1 in level $\epsilon_1, \ldots$ the degeneracies of the levels being $g_0, g_1, \ldots$ is

$$\frac{N! \, g_0^{n_0} \cdot g_1^{n_1} \cdots}{n_0! \, n_1! \cdots}$$

i.e.

$$N! \prod_i \frac{g_i^{n_i}}{n_i!},$$

the number resulting from the configuration in which n'_0 particles occupy level ϵ_0, n'_1 occupy level $\epsilon_1 \cdots$ being

$$N! \prod_i \frac{g_i^{n'_i}}{n'_i!}$$

and so on.

Equation (2.3) must therefore be amended to read

$$W_{\text{total}} = N! \prod_i \frac{g_i^{n_i}}{n_i!} + N! \prod_i \frac{g_i^{n'_i}}{n'_i!} + \cdots \tag{2.4}$$

conditions (2.1) and (2.2) remaining unchanged.

The *exact* evaluation of (2.4) is not possible, but the findings of Section 1.2 render an *almost exact* evaluation simple enough. It will be realised that each term in (2.4) refers to a different configuration, and in the last chapter we established beyond reasonable doubt firstly, that although very many configurations are possible for a macroscopic system, the number of complexions leading to one particular configuration so greatly exceeds the number leading to any configuration differing perceptibly from it that the system will always behave as though it is in its most probable configuration, and secondly, that the logarithm of the total number of complexions accessible to the system is indistinguishable for all practical purposes from the logarithm of the number of complexions leading to the most probable configuration. We can therefore replace equation (2.4) by the expression

$$\ln W_{\text{total}} \simeq \ln W_{\text{max}} = \ln N! \prod_i \frac{g_i^{n_i^*}}{n_i^*!} \tag{2.5}$$

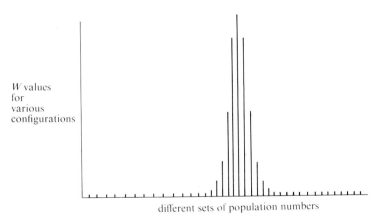

W values
for
various
configurations

different sets of population numbers

Figure 2.1

where we have denoted the population numbers leading to the most probable configuration by the symbols n_i^*. It therefore remains only to determine the set of numbers n_i^*. Figure 2.1 shows the W values of some configurations (each corresponding to a particular set of population numbers) centering around the most probable configuration. As some or all of the population numbers change, n_0 changing to $n_0 + \delta n_0$, n_1 to $n_1 + \delta n_1$ and so on, so one configuration changes to another, and it is apparent that the set of numbers corresponding to the maximum value of W must be those obtained by the solution of the equation[1]

$$\delta W = \frac{\partial W}{\partial n_0}\delta n_0 + \frac{\partial W}{\partial n_1}\delta n_1 + \cdots + \frac{\partial W}{\partial n_i}\delta n_i + \cdots = 0 \qquad (2.6)$$

the increments $\delta n_0, \delta n_1, \ldots, \delta n_i, \ldots$ being of course subject to conditions (2.1) and (2.2), which for our present purpose are contained in the equations

$$\delta E = \epsilon_0 \delta n_0 + \epsilon_1 \delta n_1 + \cdots + \epsilon_i \delta n_i \ldots \cdots = 0 \qquad (2.7)$$

and

$$\delta N = \delta n_0 + \delta n_1 + \cdots + \delta n_i + \cdots = 0. \qquad (2.8)$$

It so happens that the evaluation of any term $\partial W/\delta n_i$ is less simple than that of the term $\partial \ln W/\partial n_i$, but since when W is greatest so will be $\ln W$, the values of n_i obtained by the solution of (2.6) will be the same as those obtained by the solution of the equation

$$\delta \ln W = \frac{\partial \ln W}{\partial n_0}\delta n_0 + \frac{\partial \ln W}{\partial n_1}\delta n_1 + \cdots + \frac{\partial \ln W}{\partial n_i}\delta n_i + \cdots = 0 \quad (2.9)$$

conditions (2.7) and (2.8) remaining unchanged.

We need to combine (2.7), (2.8) and (2.9) into a single equation. It is not of course permissible to add the equations as they stand, not only because this would not be the most general way of combining them, but because the dimensions of the terms on the right hand side of each equation are not the same, those in (2.8) and (2.9) being dimensionless whereas those in (2.7) have the dimensions of energy. There is however a well-known method for combining conditions such as these, that known as Lagrange's method of undetermined multipliers. This consists of multiplying (2.8) by a dimensionless parameter α, (2.7) by a parameter $-\beta$ (we note that β must have the dimensions

[1] If y is a function of x, the maximum value of y is given by that value of x satisfying equation $dy/dx = 0$.

of inverse energy) and adding the results to (2.9) so giving the equation

$$\left(\frac{\partial \ln W}{\partial n_0} + \alpha - \beta\epsilon_0\right)\delta n_0 + \left(\frac{\partial \ln W}{\partial n_1} + \alpha - \beta\epsilon_1\right)\delta n_1 + \cdots$$
$$+ \left(\frac{\partial \ln W}{\partial n_i} + \alpha - \beta\epsilon_i\right)\delta n_i \cdots = 0 \qquad (2.10)$$

If all the increments $\delta n_0, \delta n_1, \ldots, \delta n_i, \ldots$ were *independent* and so able to assume any finite values, equation (2.10) would be satisfied by putting the contents of each bracket equal to zero, but we should not proceed quite as directly as this (although it would in fact lead to the correct set of answers) because the increments are *not* independent, but are subject to two conditions, those contained in equations (2.7) and (2.8). Only therefore if any two increments are regarded as *dependent* (the equations of dependency being (2.7) and (2.8)), may the remainder be regarded as independent. We proceed therefore as follows. We select δn_0 and δn_1 as dependent variables and assign α and β such values that

$$\frac{\partial \ln W}{\partial n_0} + \alpha - \beta\epsilon_0 = 0$$

and

$$\frac{\partial \ln W}{\partial n_1} + \alpha - \beta\epsilon_1 = 0.$$

With these values for α and β the first two terms in (2.10) vanish, and we are left with the condition

$$\left(\frac{\partial \ln W}{\partial n_2} + \alpha - \beta\epsilon_2\right)\delta n_2 + \cdots + \left(\frac{\partial \ln W}{\partial n_i} + \alpha - \beta\epsilon_i\right)\delta n_i + \cdots = 0, \qquad (2.11)$$

the remaining increments $\delta n_2, \ldots, \delta n_i, \ldots$ being now *independent* so that each is able to assume any value. Equation (2.11) is therefore satisfied only if the content of each bracket is itself equal to zero.

We see therefore that the combined condition (2.10) is satisfied only if $n_0, n_1, \ldots, n_i, \ldots$ assume such values $n_0^*, n_1^* \ldots, n_i^* \ldots$ given by the equations

$$\left.\begin{array}{c} \dfrac{\partial \ln W}{\partial n_0^*} + \alpha - \beta\epsilon_0 = 0 \\[2ex] \dfrac{\partial \ln W}{\partial n_1^*} + \alpha - \beta\epsilon_1 = 0 \\[2ex] \dfrac{\partial \ln W}{\partial n_i^*} + \alpha - \beta\epsilon_i = 0 \\[2ex] \vdots \end{array}\right\} \qquad (2.12)$$

these values for n_i^* providing the value of $\ln W_{max}$ through equation (2.5). It follows from this equation that

$$\ln W = \ln N! + \sum_i n_i^* \ln g_i - \sum_i \ln n_i^*!$$

and if we assume that each n_i^* is large enough to permit the use of the Stirling approximation formula (1.9), we may replace $\ln n_i^*!$ by $n_i^* \ln n_i^* - n_i^*$ to give the equation

$$\ln W = \ln N! + \sum_i n_i^* \ln g_i - \sum_i (n_i^* \ln n_i^* - n_i^*)$$

from which it follows that

$$\frac{\partial \ln W}{\partial n_i^*} = -\frac{\partial}{\partial n_i^*}(n_i^* \ln n_i^* - n_i^* - n_i^* \ln g_i)$$

$$= -\ln (n_i^*/g_i) \qquad (2.13)$$

Substituting into (2.12) we see that the population numbers leading to the most probable configuration[1] are those given by the equation

$$-\ln (n_i^*/g_i) + \alpha - \beta \epsilon_i = 0,$$

i.e.

$$n_i^* = g_i \exp \alpha \cdot \exp (-\beta \epsilon_i) \qquad (2.14)$$

The term $\exp \alpha$ is readily eliminated. Since $N = \sum_i n_i^*$,

$$N = \exp \alpha \{g_0 \exp (-\beta \epsilon_0) + g_1 \exp (-\beta \epsilon_1) + \cdots + g_i \exp (-\beta \epsilon_i) \ldots\}$$

so that

$$\frac{n_i^*}{N} = \frac{g_i \exp (-\beta \epsilon_i)}{\sum_i g_i \exp (-\beta \epsilon_i)} \qquad (2.15)$$

This equation is the famous Boltzmann Distribution Law.

2.4 The Particle Partition Function

We shall shortly show that the Lagrangian multiplier β is a measure of one of the properties of the *assembly*, whereas the energy levels $\epsilon_0, \epsilon_1, \ldots, \epsilon_i, \ldots$ are obviously characteristics of the particle itself. The quantity $\sum_i g_i \exp(-\beta \epsilon_i)$ appearing in equation (2.15) is therefore a *characteristic of the particle in the assembly*. It is called the *particle partition function* and is denoted by

---

[1] It follows from (2.13) that $(\partial^2 \ln W)/(\partial n_i^{*2}) = -1/n_i^*$ and is necessarily negative, so demonstrating that the values of n_i^* given by (2.14) do indeed lead to the *maximum* value for $\ln W$ and not the *minimum*.

the symbol f. It proves to be the most useful quantity we have met so far in this book, and merits some discussion before being used to obtain expressions for the properties of the assembly.

The first point which must be emphasised is that in the expression

$$f = g_0 \exp(-\beta\epsilon_0) + g_1 \exp(-\beta\epsilon_1) + \cdots + g_i \exp(-\beta\epsilon_i) + \cdots$$

we have chosen to sum over each energy level, the first term relating to level ϵ_0, the second to level ϵ_1 and so on. There is a different, but equivalent way by which f may be defined, as the sum of terms over all accessible quantum *states*, there being then g_0 identical terms relating to the g_0 states characterised by energy level ϵ_0, g_1 identical terms relating to energy level ϵ_1 and so on. In other words, whereas we have chosen to define the partition function by the equation

$$f = \sum_{\text{levels}} g_i \exp(-\beta\epsilon_i), \tag{2.16}$$

it could have been defined equally well by the equation

$$f = \sum_{\substack{\text{quantum} \\ \text{states}}} \exp(-\beta\epsilon_i), \tag{2.17}$$

its numerical value being of course the same. Some authors prefer to use (2.17) rather than (2.16). We prefer otherwise, if only for the reason that the appearance of the degeneracies in (2.17) ensures that they are not overlooked, and in this book we shall always sum over energy levels (and this will always be implied by the symbol $\sum_i$) unless otherwise stated.

The second point to be made is that the quantity f defined by (2.16) or (2.17) is not the only quantity in statistical mechanics which is called a partition function. The quantity f is the partition function for the *particle*, and has meaning only for an assembly of *independent* particles. In Chapter 8, when studying a closed macroscopic system in equilibrium with a heat bath we shall make use of a quantity Q which is the partition function for the *system*, and in Chapter 12 when studying an open macroscopic system permitted to exchange both energy and particles with the surroundings we shall make use of a third quantity which is the *grand* partition function for the system.

2.5   The Thermodynamic Properties of the Assembly

We are at last in a position to obtain useful expressions for the thermodynamic properties of an assembly of N independent, localised particles at constant E and V. Denoting the most probable distribution numbers by the symbols $n_0, n_1, \ldots, n_i, \ldots$ instead of $n_0^*, n_1^*, \ldots, n_i^*, \ldots$ since no confusion is now possible, we first express the Boltzmann Distribution Law rather more economically by the equation

$$n_i = \frac{N}{f} g_i \exp(-\beta \epsilon_i) \qquad (2.18)$$

and then describe the assembly by the fundamental equations

$$N = \sum_i n_i \qquad (2.19)$$

$$E = \sum_i n_i \epsilon_i$$

$$= \frac{N}{f} \sum_i \epsilon_i g_i \exp(-\beta \epsilon_i) \qquad (2.20)$$

and

$$S = k \ln W_{\text{total}} \simeq k \ln W_{\text{max}}$$

$$= k \ln \left[N! \prod_i \frac{g_i^{n_i}}{n_i!} \right] \qquad (2.21)$$

Rather more useful expressions than (2.20) and (2.21) can be obtained. We consider first that for the entropy of the assembly.

By making use of the Stirling approximation formula we first express (2.21) in the form

$$\frac{S}{k} = N \ln N - N + \sum_i n_i \ln g_i - \sum_i (n_i \ln n_i - n_i)$$

$$= N \ln N - \sum_i \left(n_i \ln \frac{n_i}{g_i} \right).$$

From (2.18)

$$\ln \frac{n_i}{g_i} = \ln \frac{N}{f} - \beta \epsilon_i$$

so that

$$\sum_i \left(n_i \ln \frac{n_i}{g_i} \right) = N \ln \frac{N}{f} - \beta \sum_i (n_i \epsilon_i).$$

It follows that

$$\frac{S}{k} = N \ln f + \beta E. \qquad (2.22)$$

Before obtaining expressions for the other thermodynamic functions it is profitable to determine the quantity $(\partial S/\partial E)_{V,N}$. From (2.22)

$$\left(\frac{\partial S}{\partial E}\right)_{V,N} = Nk\left(\frac{\partial \ln f}{\partial E}\right)_{V,N} + \beta k + kE\left(\frac{\partial \beta}{\partial E}\right)_{V,N}. \qquad (2.23)$$

Now

$$Nk\left(\frac{\partial \ln f}{\partial E}\right)_{V,N} = \frac{Nk}{f}\left(\frac{\partial f}{\partial \beta}\right)_{V,N}\left(\frac{\partial \beta}{\partial E}\right)_{V,N}$$

$$= \frac{Nk}{f}\left(\frac{\partial \beta}{\partial E}\right)_{V,N}\frac{\partial}{\partial \beta}\left\{\sum_i g_i \exp\left(-\beta\epsilon_i\right)\right\}$$

$$= \frac{Nk}{f}\left(\frac{\partial \beta}{\partial E}\right)_{V,N}\sum_i (-\epsilon_i)g_i \exp\left(-\beta\epsilon_i\right)$$

$$= k\left(\frac{\partial \beta}{\partial E}\right)_{V,N}\left\{-\sum_i (n_i\epsilon_i)\right\}$$

$$= -kE\left(\frac{\partial \beta}{\partial E}\right)_{V,N} \qquad (2.24)^1$$

It follows that the first and third terms on the right-hand side of (2.23) cancel, and we are left with the equation

$$\left(\frac{\partial S}{\partial E}\right)_{V,N} = \beta k. \qquad (2.25)$$

But classical thermodynamics shows that for a closed system in equilibrium

$$\left(\frac{\partial S}{\partial E}\right)_{V,N} = \frac{1}{T}$$

so that

$$\beta = \frac{1}{kT}. \qquad (2.26)$$

We see therefore that β, introduced as a Lagrangian multiplier, acquires a physical meaning, being the statistical–mechanical measure of the thermodynamic temperature of the assembly. We can from now on express any equations in terms of β or in terms of T whichever is the more convenient.

---

[1] See Exercise 2.1.

We will adopt the latter procedure for the time being, and write (2.22) in the form

$$S = kN \ln f + \frac{E}{T}. \tag{2.27}$$

The most useful expression for the energy of the assembly is obtained as follows: from (2.16) and (2.26)

$$f = \sum_i g_i \exp\left(-\frac{\epsilon_i}{kT}\right). \tag{2.28}$$

Hence

$$\left(\frac{\partial f}{\partial T}\right)_V = \sum_i \left\{\frac{\epsilon_i}{kT^2} g_i \exp\left(-\frac{\epsilon_i}{kT}\right)\right\}$$

so that

$$\frac{kNT^2}{f}\left(\frac{\partial f}{\partial T}\right)_V = \sum_i \left\{\epsilon_i \left[\frac{N}{f} g_i \exp\left(-\frac{\epsilon_i}{kT}\right)\right]\right\}.$$

It follows from (2.18) that the quantity in square brackets is n_i, and since $(1/f)\,\partial f/\partial T = \partial \ln f/\partial T$,

$$kNT^2\left(\frac{\partial \ln f}{\partial T}\right)_V = \sum_i (n_i \epsilon_i) = E. \tag{2.29}$$

Substituting (2.29) into (2.27) we obtain the equation

$$S = kN \ln f + kNT\left(\frac{\partial \ln f}{\partial T}\right)_V \tag{2.30}$$

which is, in many instances, rather more easily handled than (2.27). The statistical–mechanical expression for the Helmholtz function A (defined in classical thermodynamics by the equation $A = E - TS$) follows from (2.29) and (2.30), i.e.

$$A = -kNT \ln f, \tag{2.31}$$

whilst those for the heat capacity C_v, the pressure P and the chemical potential μ can be obtained from the classical relationships

$$C_v = \left(\frac{\partial E}{\partial T}\right)_{V,N} \tag{2.32}$$

$$P = -\left(\frac{\partial A}{\partial V}\right)_{T,N} \tag{2.33}$$

$$\mu = \left(\frac{\partial A}{\partial N}\right)_{T,V}. \tag{2.34}$$

EXERCISES

2.1 An important consequence of the distribution law is that so long as the energy levels accessible to a particle are unchanged

$$N\frac{\partial \ln f}{\partial X} = -E\frac{\partial \beta}{\partial X} \qquad (2.E.1)$$

where X is any variable. It will be shown in the next chapter that the energy levels accessible to a localised particle are determined by the volume it occupies (i.e. V/N) and that those accessible to a non-localised particle are determined by the total volume available to it, (i.e. V), so that (2.E.1) should be properly expressed as

$$N\left(\frac{\partial \ln f}{\partial X}\right)_{V/N} = -E\left(\frac{\partial \beta}{\partial X}\right)_{V/N} \qquad (2.E.2)$$

for an assembly of independent localised particles, and as

$$N\left(\frac{\partial \ln f}{\partial X}\right)_{V} = -E\left(\frac{\partial \beta}{\partial X}\right)_{V} \qquad (2.E.3)$$

for an assembly of independent non-localised particles.

Prove equation (2.E.2).

(It will be recalled that this relation has already been used in the derivation of equation (2.24). It will be required again for the establishment of equation (4.31) in Chapter 4, and used again in Section 7.1).

CHAPTER 3

Atomic Crystals:
The Einstein Treatment
and the Debye Modification

Although we said at the beginning of Chapter 2 that our aim was to construct a model through which we would obtain understanding of the behaviour of an atomic crystal, the formulae obtained so far are quite general in the sense that they apply to assemblies of independent localised particles whatever their complexity. In this chapter we revert to our original aim and suppose that the particles in the model represent atoms in a crystal, and our principal task therefore is to discover the relationship between the various energy levels $\epsilon_0, \epsilon_1, \ldots, \epsilon_i, \ldots$ accessible to such atoms, so as to obtain, through the partition function, theoretical expressions for the properties of the crystal for comparison with the corresponding experimental values.

3.1 Nuclear and Electronic States

Expressions for the most important contributions to the properties of an atomic crystal are obtained by treating the atoms as mass points capable only of vibrational motion. An atom is not however a mass point, but has a nuclear and electronic sub-structure, and is capable, in principle at least, of existing in more than one nuclear state and in more than one electronic state. We shall consider matters such as these in greater detail in Chapters 4 and 5 when studying the behaviour of non-localised molecules, and the following remarks are presented therefore in this section without proof. Firstly, the spacings between different nuclear energy levels are so great that the chance that an atom is found in any but a nuclear ground state is vanishingly small. Secondly, that although the degeneracy of the nuclear ground energy level would indeed contribute to the absolute value of some properties of the system, such as its entropy, the inclusion of such terms is without practical significance because the nuclear ground states of atoms are unchanged as the result of their involvement in chemical reactions or phase changes, and such contributions would therefore cancel out. The whole matter of nuclear states can therefore be ignored. The question of electronic states is similar in some respects but not all. The spacings between different electronic energy levels, other than for a few diatomic molecules which are not our present concern, are so great that we may suppose without sensible error that all atoms in a crystal remain in their electronic ground states during all processes in which we are interested, *but* the degeneracies of the ground electronic energy levels are *not* necessarily conserved during phase changes and chemical reactions and must therefore be included in the relevant formulae.

All this means, and here again we postpone justification until Chapters 4 and 5, that the partition function figuring in the formulae of the last chapter, when applied to atoms in a crystal, depends only on the electronic ground state degeneracy denoted by the symbol g_{e_0}, and the vibrational energy states accessible to the atom, and is given by the equation

$$f = g_{e_0} f_{\text{vib}} \qquad (3.1)$$

where f_{vib} is the vibrational partition function defined by the equation

$$f_{\text{vib}} = \sum_i g_i \exp\left(-\epsilon_i/kT\right) \qquad (3.2)$$

where the symbols ϵ_i and g_i now denote the vibrational energy levels and their degeneracies.

3.2   The Vibrational Partition Function: The Einstein Treatment

The first attempt to determine the relationship between the various vibrational energy levels in a crystal was made by Einstein in 1907. In his treatment it is assumed that the vibrational modes of an atom may be likened to those of a simple harmonic oscillator, so that it is to the mechanics of such an oscillator that we now turn.

Let us suppose that a number of point masses are arranged in a straight line and that each is capable of limited movement so that, due to the forces between them, each vibrates about its mean position. The situation is portrayed in Figure 3.1. Figure 3.1(a) shows a number of point masses in their mean positions, and Figure 3.1(b) shows one point mass displaced a distance d from its mean position. If we suppose that the vibrations are subject to Hooke's law (that is to say, the restoring force operating on any one point mass is proportional to its displacement) the dependence of the potential energy of each point mass on its displacement will be as shown in Figure 3.1(c).

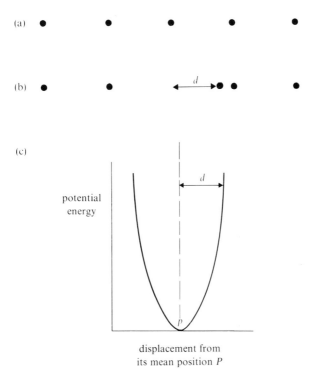

Figure 3.1   (a) All point masses shown in their mean positions.
                (b) One point mass shown displaced a distance *d*.
                (c) The dependence of the potential energy of a point mass on its displacement.

Such a vibration as that described is said to be *simple harmonic*, and if the point mass is subject only to forces acting in a single direction (as we have supposed above) each point mass is said to be a *one-dimensional* simple harmonic oscillator.

Quantum theory shows that the energy levels of such an oscillator are given by the equation

$$\epsilon_i = (i + \tfrac{1}{2})\frac{h}{2\pi}\left(\frac{\lambda}{m}\right)^{\frac{1}{2}} \tag{3.3}$$

where h is Planck's constant, m is the mass of the oscillator, λ the force constant governing the vibration, and i the vibrational quantum number taking the values $0, 1, 2, \ldots$.

The quantity $(1/2\pi)(\lambda/m)^{\frac{1}{2}}$ has the dimensions of frequency and is denoted by the symbol v, so that the permitted energy levels can be written more concisely as

$$\epsilon_i = (i + \tfrac{1}{2})hv. \tag{3.4}$$

These energy levels are non-degenerate, so that the partition function for the one-dimensional oscillator (written f_{odo}) is easily framed:

$$f_{\text{odo}} = \exp\left(-\frac{hv}{2kT}\right) + \exp\left(-\frac{3hv}{2kT}\right) + \exp\left(-\frac{5hv}{2kT}\right) + \cdots$$

$$= \exp\left(-\frac{hv}{2kT}\right)\left[1 + \exp\left(-\frac{hv}{kT}\right) + \exp\left(-\frac{2hv}{kT}\right) + \cdots\right].$$

The expression in brackets may be written as $1 + x + x^2 + \cdots$ where $x = \exp(-hv/kT)$, and since x is necessarily smaller than unity, and in these circumstances

$$1 + x + x^2 + \cdots = \frac{1}{1 - x},$$

the expression for the partition function becomes

$$f_{\text{odo}} = \frac{\exp(-hv/2kT)}{1 - \exp(-hv/kT)}. \tag{3.5}$$

It is convenient to introduce a parameter θ by the equation

$$\theta = \frac{hv}{k} = \frac{h}{2\pi k}\left(\frac{\lambda}{m}\right)^{\frac{1}{2}} \tag{3.6}$$

(the parameter θ has the dimensions of temperature and is called the characteristic temperature for the oscillator) and so write

$$f_{\text{odo}} = \frac{\exp(-\theta/2T)}{1 - \exp(-\theta/T)}. \tag{3.7}$$

Quantum theory shows also that the energy levels of an isotropic (i.e. spherically symmetrical) three-dimensional oscillator are given by the equation

$$\epsilon_i = (i + \tfrac{3}{2})\frac{h}{2\pi}\left(\frac{\lambda}{m}\right)^{\frac{1}{2}} = (i + \tfrac{3}{2})hv \tag{3.8}$$

where each symbol has the same meaning as before, each energy level being $\frac{1}{2}(i + 1)(i + 2)$-fold degenerate. (The origin of this degeneracy is easily seen.

The vibration may be resolved into three components directed along three mutually-perpendicular axes. The degeneracy of an energy level i is the number of distinguishable ways in which i quanta may be distributed in those three directions. When i is zero, g_i is unity. When i is one, g_i is *three* because it can operate in any of of three directions. When i is two we have *six* possible distributions 200, 020, 002, 110, 011 and 101. When i is three we have the *ten* possible distributions 300, 030, 003, 210, 021, 201, 120, 012, 102 and 111. The formula $\frac{1}{2}(i + 1)(i + 2)$ is merely the general term for the series 1, 3, 6, 10, ...)

The partition function for the three-dimensional oscillator, (denoted by f_{tdo}) is therefore given by the equation

$$f_{tdo} = \exp\left(-\frac{3hv}{2kT}\right) + 3\exp\left(-\frac{5hv}{2kT}\right) + 6\exp\left(-\frac{7hv}{2kT}\right)$$

$$+ 10\exp\left(-\frac{9hv}{2kT}\right) + \cdots$$

$$= \frac{\exp\left(-3hv/2kT\right)}{[1 - \exp\left(-hv/kT\right)]^3}$$

$$= \left[\frac{\exp\left(-hv/2kT\right)}{1 - \exp\left(-hv/kT\right)}\right]^3 \tag{3.9}$$

from which we see that

$$f_{tdo} = f_{odo}^3. \tag{3.10}$$

We again introduce the parameter θ which has the same significance as before and so obtain the final equation

$$f_{tdo} = \frac{\exp\left(-3\theta/2T\right)}{[1 - \exp\left(-\theta/T\right)]^3}. \tag{3.11}$$

The fundamental assumptions of the Einstein treatment are that the vibrational behaviour of an assembly of N independent localised atoms is equivalent to that of N three-dimensional simple harmonic oscillators (or if we prefer, to that of $3N$ one-dimensional oscillators), and that *all vibrations are governed by the same value for the force constant*. In effect therefore, the Einstein assumption is that all vibrational modes of an atomic crystal are governed by a single parameter θ (given by equation 3.6) and that the vibrational partition function for the atom is given by the equation

$$f_{vib} = \frac{\exp\left(-3\theta/2T\right)}{[1 - \exp\left(-\theta/T\right)]^3}. \tag{3.12}$$

It then follows that

$$\ln f_{vib} = -3\theta/2T - 3\ln\left[1 - \exp\left(-\theta/T\right)\right], \tag{3.13}$$

and

$$\left(\frac{\partial \ln f_{\text{vib}}}{\partial T}\right)_{V/N} = 3\theta/2T^2 + \frac{3\theta/T^2}{\exp(\theta/T) - 1}. \tag{3.14}$$

Before proceeding an explanation must be given for imposing constancy of V/N in (3.14). It is reasonable to suppose that the force constant governing the vibration of a one-dimensional oscillator depends on the distance between the point masses, and therefore that that governing the vibration of a three-dimensional oscillator depends on the volume to which the oscillator is confined. In the case of the atom this is of course simply the atomic volume V/N. It follows from (3.6), (3.8) and (3.12) that the fundamental frequency v, the permissible energies ϵ_i, the characteristic temperature θ and the partition function for the atom are all functions of V/N. We shall find in Section 3.4 that it is this dependence that leads to a positive (non-zero) value for the pressure of a crystal, the pressure being given by the equation

$$P = NkT\left(\frac{\partial \ln f}{\partial V}\right)_{T,N},$$

and, when we produce an expression for the chemical potential of the crystal, for the small but non-zero value for the quantity $(\partial \ln f/\partial N)_{V,T}$.

Expressions for the thermodynamic properties of an atomic crystal may now be obtained from the equations derived in the last chapter, and by making use of (3.1), (3.13) and (3.14). We shall give them only for A, S, E and the heat capacity C_v, but shall give an approximate expression for μ, the chemical potential in Section 3.4.

$$A = -NkT \ln f$$

$$= -NkT \ln g_{e_0} + \tfrac{3}{2}Nk\theta + 3NkT \ln\left[1 - \exp\left(-\frac{\theta}{T}\right)\right] \tag{3.15}$$

$$S = Nk \ln f + NkT\left(\frac{\partial \ln f}{\partial T}\right)_{V/N}$$

$$= Nk \ln g_{e_0} + 3Nk\left[\frac{\theta/T}{\exp(\theta/T) - 1} - \ln\left\{1 - \exp\left(-\frac{\theta}{T}\right)\right\}\right] \tag{3.16}$$

$$E = NkT^2(\partial \ln f/\partial T)_{V/N}$$

$$= \tfrac{3}{2}Nk\theta + \frac{3Nk\theta}{\exp(\theta/T) - 1}, \tag{3.17}$$

$$C_v = (\partial E/\partial T)_v$$

$$= 3Nk\left(\frac{\theta}{T}\right)^2\left\{\frac{\exp(\theta/T)}{[\exp(\theta/T) - 1]^2}\right\}. \tag{3.18}$$

It will be noted that the term in g_{e_0} occurs only in the expressions for A and S. Expressions for the *molar* properties for the crystal may of course be obtained by replacing N by L the Avogadro constant.[1]

Inspection of the right-hand side of (3.17) shows that the magnitude of the second term increases as T increases, but approaches zero as $T \to 0$, whereas the magnitude of the first term (which appears also in (3.15)) is independent of temperature and must therefore represent the vibrational energy of the crystal at absolute zero when all atoms are in their lowest possible vibrational state, $(i = 0)$, the quantity $3k\theta/2$ being the so-called zero-point energy of the atom. The prediction that a crystal retains a non-zero vibrational energy at absolute zero is a consequence of quantum theory and is quite foreign to classical mechanics.

Equation (3.18) is of particular interest because, once a value for θ is obtained from the experimental value for C_v at one temperature, 'theoretical' values for C_v over a range of temperatures may be obtained for comparison with experimental values, so providing a test for both the applicability of the statistical–mechanical equations obtained in Chapter 2, and of their interpretation by means of the Einstein hypothesis. The graph of C_v against T/θ is shown in Figure 3.2, using the data in Table 3.1.

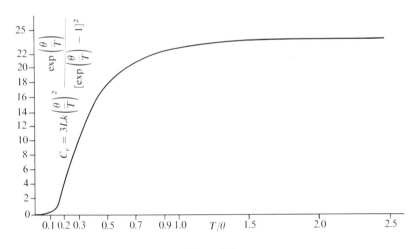

Figure 3.2

[1] The quantities

$$\left(\frac{\theta}{T}\right)^2 \frac{\exp(\theta/T)}{[\exp(\theta/T) - 1]^2}, \quad \frac{\theta/T}{\exp(\theta/T) - 1} \quad \text{and} \quad -\ln[1 - \exp(-\theta/T)]$$

are known as the Einstein functions. Their values for many values of θ/T are given in the literature: see for example J. Sherman and R. B. Ewell *J. Phys. Chem.* **46,** 641 (1942).

Table 3.1   Predicted values for the molar heat capacity C_v/JK^{-1} mol^{-1} for various values of the parameter θ/T

θ/T	$C_v/\text{JK}^{-1}\,\text{mol}^{-1}$	θ/T	$C_v/\text{JK}^{-1}\,\text{mol}^{-1}$
10.8	0.058	2.0	18.10
10.0	0.112	1.8	19.20
9.0	0.250	1.6	20.27
8.0	0.537	1.4	21.27
7.0	1.125	1.2	22.15
6.0	2.25	1.0	23.00
5.0	4.27	0.8	23.70
4.0	7.60	0.6	24.25
3.0	12.40	0.4	24.65

The predicted values for C_v in two limiting cases are of particular interest, first that when the temperature is so low that θ/T is very much greater than unity, and second when the temperature is so high that θ/T approaches zero. Since $\exp(y) = 1 + y + y^2/2! + y^3/3! \ldots$ the quantity in braces in (3.18) may be expanded to give

$$\frac{1 + \theta/T + \theta^2/2T^2 + \theta^3/6T^3 + \cdots}{[\theta/T + \theta^2/2T^2 + \theta^3/6T^3 + \cdots]^2}.$$

When $\theta/T \gg 1$, we may ignore unity in the numerator, so that the quotient reduces to $(\theta/T + \theta^2/2T^2 + \theta^3/6T^3 + \cdots)^{-1}$ and so to zero as T tends to zero, so that

$$C_v \to 0 \quad \text{as} \quad T \to 0, \tag{3.19}$$

and at temperatures so high that $\theta/T \ll 1$, all terms except unity in the numerator may be ignored as may all terms except θ/T in the denominator, so that the quantity in braces approaches $(T/\theta)^2$, and C_v approaches the value $3Nk$. Thus (3.18) predicts that at high temperatures the *molar* heat capacity approaches $3Lk$, i.e.

$$C_v \to 3Lk \quad \text{i.e. 3 R, i.e. } 24.94\,\text{JK}^{-1}\,\text{mol}^{-1}, \tag{3.20}$$

both conclusions being reflected in Figure 3.2.

These two predictions are very impressive indeed. The experimental fact, established in the nineteenth century, that the heat capacity of a monatomic crystalline body (or indeed that of *any* crystalline body) approaches zero at very low temperatures was quite inexplicable in terms of classical mechanics (and was one of the great unsolved problems of classical physics), and the fact that quantum theory predicts just this behaviour, as demonstrated above, proved to be a very powerful argument for the acceptance of the theory of quantisation of energy. The prediction that C_v approaches $3Lk$ only at temperatures so high that $\theta/T \ll 1$ is equally impressive. In 1819 Dulong and Petit found experimentally that at room temperature the atomic heats (i.e. the product of the gramme atomic weights and the specific heats) of many solid elements are very nearly the same and equal to about 6 cal K^{-1} i.e. about 25 JK^{-1}, which is of course in precise agreement with (3.20). But certain solids, in particular diamond, boron and beryllium, were found to be exceptional in showing values much below these figures: thus diamond shows the value 5.65 $JK^{-1}\,mol^{-1}$, boron 10.46 $JK^{-1}\,mol^{-1}$ and beryllium 14.6 $JK^{-1}\,mol^{-1}$. These exceptions to the Dulong and Petit law are also inexplicable in terms of classical physics (which predicts a value of $3Lk$ at *all* temperatures) but are rationalised by Figure 3.2 which shows that those elements which obey the law at room temperature, say 300 K, are those for which θ is less than about 200 K, so that T/θ is greater than 1.5, whilst those which at room temperature show values for C_v very much smaller than 25 $JK^{-1}\,mol^{-1}$ are those for which θ is much greater than 300 K, so that at room temperature T/θ is much smaller than unity.

The attraction of the Einstein treatment is that it is possible by considering the physical characteristics of the solid to predict whether it is likely to obey the Dulong and Petit law or not. It will be recalled that

$$\theta = \frac{h\nu}{k} = \frac{h}{2\pi k}\left(\frac{\lambda}{m}\right)^{\frac{1}{2}} \tag{3.6}$$

where m is the mass of the oscillator and λ the force constant governing its vibration, so applying this formula to the vibration of an atom in a solid, we would expect to find that θ is exceptionally high for those elements for which m is small and λ exceptionally large. What are the factors that determine the value of the force constant? It is reasonable to suppose that λ will be exceptionally high if each atom is bound exceptionally strongly to its neighbours, and the stronger the binding the higher would we expect to find its melting point, the smaller its atomic volume V/N, the *harder* the solid (the hardness of a body being the same as the degree of difficulty experienced in distorting it) and the smaller its coefficient of compressibility $-V^{-1}(\partial V/\partial P)_T$. There have been several attempts to predict values for λ, ν or θ for an atomic solid from its physical characteristics. One, due originally to Einstein[1] gives the connection between θ and the elastic constants (Poisson's ratio and the coefficient of compressibility), and a simpler (and earlier) formula due to

---

[1] The Einstein formula is given by Fowler, *Statistical Mechanics*, (Cambridge University Press, Paperback Edition page 127).

Lindemann[1] connects v with the melting point of the solid and its molar atomic volume. This formula

$$v = \frac{A}{V_m^{\frac{1}{3}}}\left(\frac{T_m}{M}\right)^{\frac{1}{2}}$$

in which V_m is the molar atomic volume, T_m the melting point and M the (dimensionless) atomic weight is semi-empirical in the sense that the value for the constant A was obtained by correlating for various atomic solids values for v estimated by other means with those for V_m, T_m and M. If V_m is expressed in cm³, A has the value 2.8×10^{12} cm $K^{-\frac{1}{2}} s^{-1}$.

The atoms in both diamond and boron are bound to their neighbours by very strong covalent bonds, and they are both elements of very low atomic weight, so that λ would be expected to be high and m small. It is not surprising therefore that θ for both solids is exceptionally high. The case of beryllium is different in so far that the bonds between the atoms in the crystal are only partially covalent (so that we would expect the value of λ to be considerably smaller than the corresponding values for diamond and boron), and therefore we must suppose that the moderately-large value for θ for beryllium depends more on the smallness of m than on the magnitude of the force constant. We have all the information available to make a very rough estimate of the values for θ, v and λ for each of these elements from thermodynamic data. It is clear that if Figure 3.2 *truly* represents the variation of C_v with temperature, the values for θ, v and λ for any element may be estimated if an experimental value for C_v (falling between 2 and 20 JK^{-1} mol^{-1}) at some one temperature is known. Thus the value for C_v for beryllium at 300 K is 14.6 JK^{-1} mol^{-1}. It appears from Figure 3.2 that the corresponding value for T/θ is 0.39, so that $\theta = 769$ K, and

$$v = \frac{\theta k}{h} = 16 \times 10^{12} \ s^{-1}.$$

The atomic weight of beryllium is 9.01 so that

$$m = 0.00901/(6.023 \times 10^{23}) \ kg,$$

so that

$$\lambda = 4\pi^2 v^2 m = 151 \ Nm^{-1}.$$

These figures and the results of similar calculations for diamond and boron are shown in Table 3.2. They are as expected from the argument given above.

The apparent success of the Einstein treatment in accounting for the thermal behaviour of solid elements is impressive, but a more detailed investigation shows that the agreement is qualitative rather than exact.

---

[1] Lindemann, *Physikal. Z.* **11**, 609, 1910.

Table 3.2   Characteristics of elements estimated from the experimental values for C_v at 300 K

	M	$\dfrac{C_V}{\text{JK}^{-1}\,\text{mol}^{-1}}$	T/θ	θ/K	$v \times 10^{-12}\,\text{s}$	λ/Nm^{-1}
diamond	12	5.65	0.22	1364	28.4	635
boron	10.5	10.46	0.30	1000	20.8	307
beryllium	9	14.6	0.39	769	16.0	151

The failure of the treatment is illustrated by the following. If an experimental value for C_v at room temperature is used to estimate θ, as shown above, and this value used to predict values for C_v at other temperatures using equation (3.18) it is found that although the predicted values at high temperatures are reasonably satisfactory, those at very low temperatures show a systematic divergence from experimental values, the calculated values being smaller than those obtained by experiment. In other words, the curve in Figure 3.2 falls away rather too rapidly at low temperatures. Furthermore, the predicted values fail to obey the experimentally-established law that at *very* low temperatures (5 to 15 K) the values for C_v are proportional to the third power of the temperature. The failure of the Einstein treatment is illustrated even more vividly if, from the estimated value for θ (obtained as shown above) the value for the molar entropy is calculated using equation (3.16). Thus using the value $\theta = 769$ K, the calculated value for the molar entropy of beryllium at 300 K is 7.36 JK^{-1} mol^{-1}, whereas the experimental value is 11.09 JK^{-1} mol^{-1}. Agreement is somewhat better, but still quite unsatisfactory, even for those elements for which θ is low.

The fault (as regards the values of C_v) lies almost completely in the Einstein assumption that for a crystal containing N atoms it is sufficiently accurate to take all $3N$ frequencies equal.

3.3   The Debye Modification

A modification of the Einstein treatment was developed by Debye in 1912 based on the replacement of the Einstein assumption by the theory that $3N$ frequencies could be taken as the $3N$ *lowest* frequencies of a continuum with the same elastic properties as the crystal. In the Debye theory the atoms are assumed therefore to have a range of vibrational frequencies bounded by a maximum frequency v_{max}, and it is shown that the properties of the crystal depend on the parameters

$$u = \theta_D/T \tag{3.21}$$

and

$$\theta_D = h v_{max}/k. \tag{3.22}$$

The formula for the molar thermal capacity then becomes

$$C_v = 2Lk\left[4D - \frac{3u}{\exp{(u)} - 1}\right] \tag{3.23}$$

where

$$D = \frac{3}{u^3}\int_0^u \frac{X^3\,dX}{\exp{(X)} - 1} \tag{3.24}$$

and

$$X = \frac{hv}{kT}. \tag{3.25}$$

The Debye theory predicts that at high temperatures C_v approaches the Dulong and Petit value $3Lk$, and that at very low temperatures

$$C_v \rightarrow \frac{12Lk\pi^4}{5}\left(\frac{T}{\theta_D}\right)^3 \tag{3.26}$$

so that $C_v \propto T^3$ as experimental results require. Perhaps the most impressive evidence for the Debye modification is that if θ_D is obtained for any monatomic solid from an experimental value for C_v at one temperature (using a similar method to that described in the last section) and experimental values for C_v for each temperature plotted against T/θ_D, the data for all such solids falls on a common curve. In other words, the value for C_v for element 1 at temperature T_1 is the same as that for element 2 at temperature T_2 if

$$\frac{T_1}{\theta_{D_1}} = \frac{T_2}{\theta_{D_2}}. \tag{3.27}$$

This is an example of the so-called *law of corresponding states*.

The fact that the Debye modification is very successful in predicting the behaviour of atomic crystals at very low temperatures must not be allowed to detract from the worth of the Einstein treatment from which, of course, it derives. For predicting the properties of monatomic solids at high temperatures there is, in any event, little to choose between them (they are both 'approximate' theories) but perhaps more important is the point that it is through the Einstein treatment (by virtue of its extreme *simplicity*) that we so easily gain insight into the behaviour of atomic crystals. This is the very essence of statistical mechanics—at least, as taught in this book.

3.4 Statistical–Mechanical Expressions for the Pressure and Chemical Potential of an Atomic Crystal

Both topics discussed in this section illustrate the fact that the partition function of an atom in a crystal is a function of the atomic volume V/N.

The pressure of any closed system is related to its Helmholtz function by the equation

$$P = -\left(\frac{\partial A}{\partial V}\right)_{T,N} \tag{3.28}$$

and is therefore given by the statistical–mechanical expression

$$P = NkT\left(\frac{\partial \ln f}{\partial V}\right)_{T,N}. \tag{3.29}$$

We shall now demonstrate that the Einstein treatment leads to a positive (non-zero) value for the pressure. Since

$$\ln f = \ln g_{e_0} - \frac{3\theta}{2T} - 3\ln\left[1 - \exp\left(-\theta/T\right)\right]$$

it follows that

$$\left(\frac{\partial \ln f}{\partial V}\right)_{T,N} = -\frac{3}{T}\left[\frac{1}{2} + \frac{1}{\exp\left(\theta/T\right) - 1}\right]\left(\frac{\partial \theta}{\partial V}\right)_{T,N} \tag{3.30}$$

the quantity in square brackets being positive in all circumstances. We earlier showed that increase in the atomic volume V/N leads to a decrease in the value of θ. Increase in the total volume V, N remaining constant increases V/N so that $(\partial\theta/\partial V)_{T,N}$ is negative. It follows that $(\partial \ln f/\partial V)_{T,N}$ and hence the pressure is necessarily positive.

The chemical potential of a one-component system may be defined by either of the equations

$$\mu = \left(\frac{\partial A}{\partial N}\right)_{T,V}, \tag{3.31}$$

or

$$\mu = \frac{G}{N} \tag{3.32}$$

where G is the Gibbs function $A + PV$. It follows from equations (2.31) and (3.31) that a formally-correct expression for μ is

$$\mu = -kT \ln f - NkT \left(\frac{\partial \ln f}{\partial N} \right)_{V,T}$$

$$= -kT \ln f + \frac{V}{N} kT \left(\frac{\partial \ln f}{\partial (V/N)} \right)_{T}. \tag{3.33}$$

It is however recognised in classical thermodynamics that the magnitudes of the Helmholtz and Gibbs functions for condensed systems are so close that we are justified in writing

$$A \simeq G \tag{3.34}$$

and so,

$$G \simeq -NkT \ln f$$

and so (from 3.32)

$$\mu = \frac{G}{N} \simeq -kT \ln f \tag{3.35}$$

from which it appears that the second term on the right-hand side of (3.33) is negligible compared with the first. The (approximate) expression for μ in terms of the Einstein parameter θ, and taking into account the possible degeneracy of the electronic ground state follows from (3.35). It is

$$\mu \simeq -kT \ln g_{e_0} + \frac{3k\theta}{2} + 3kT \ln [1 - \exp(-\theta/T)]. \tag{3.36}$$

3.5   Absolute Energies and Energy Zeros

Although (assuming that the crystal is adequately described by the Einstein treatment) an *absolute* value for the vibrational energy of a crystal is given by equation (3.17), and although we might formally include terms for the nuclear and electronic ground state energies (to neither of which can we assign numerical values) we must not suppose that the resulting expression

$$E = E_{\text{vib}} + E_n + E_e$$

gives the *total* energy in an absolute sense. It will of course be realised that an expression such as 'the absolute value for the energy of a system' is without physical significance because such values are unobtainable, and when we speak of the energy associated with a body when in a particular state, we really mean the difference between that energy and that with which it would be associated were it in some 'standard state' which we arbitrarily and conventionally 'define' as being of zero energy, just as when we speak of a mountain as being 15 000 feet high, we really mean that its summit is 15 000 feet above sea-level. It will also be realised that the fact that we cannot assign absolute values to the energy (or Helmholtz or Gibbs functions) of a system is of no importance, because we are interested only in the change in the values of these properties as the result of a physical or chemical process, and all that we must ensure is that we employ the same 'energy zero' throughout the exercise.

The energy E figuring in equation (3.17) is the vibrational energy of the assembly relative to that unrealisable state in which the particles occupy their equilibrium positions in the crystal but possess *no* vibrational energy. The value of A in (3.15) has the same significance. For some purposes it is convenient to set our arbitrary zero *not* at this unrealisable state, but at the (presumably real) state in which all atoms in the crystal are in their lowest vibrational state $(i = 0)$, when the expressions for the Helmholtz function and energy become

$$A' = A - E_{0_{vib}}$$
$$= -NkT \ln g_{e_0} + 3NkT \ln [1 - \exp(-\theta/T)] \qquad (3.37)$$

and

$$E' = E - E_{0_{vib}},$$
$$= \frac{3Nk\theta}{\exp(\theta/T) - 1} \qquad (3.38)$$

where we have written $\frac{3}{2}Nk\theta$ as $E_{0_{vib}}$, this quantity being the vibrational energy at absolute zero.

If we wish to compare the energy (or the Helmholtz function) of a crystal in one state (say at one temperature) with that of the same crystal in another state (say at another temperature) it is immaterial whether we use expressions for E and A or for E' and A' (as long as we are consistent), but in certain situations (when studying multi-phase systems or chemical reactions) neither convention is suitable. If, for example, we wish to compare the molar energy of a crystal with that of a gaseous phase in equilibrium with it, we must ensure that the energy zero implicit in the expression for the energy of one is the same as that implicit in the expression for the energy of the other.

A perfect gas is one, by definition, in which no potential energies exist between the individual molecules, whereas in the case of the atomic crystal potential energy terms exist between the atoms by virtue of their position. When comparing the energies of gas and crystal we must therefore 'remove' the potential energy terms implicit in the expressions for the energy of the latter.

This is achieved by expressing the possible energy levels accessible to the atoms in the crystal as $(e + \epsilon_0), (e + \epsilon_1), \ldots, (e + \epsilon_i), \ldots$ where e (a negative quantity) is the potential energy of the atom in the crystal relative to that at infinite separation, whereas $\epsilon_0, \epsilon_1, \epsilon_i, \ldots$ have the same significance as heretofore. The partition function now becomes

$$f'' = g_{e_0} \cdot \left[g_0 \exp - \frac{e + \epsilon_0}{kT} + g_1 \exp - \frac{e + \epsilon_1}{kT} + \cdots \right]$$

$$= g_{e_0} \cdot \exp(-e/kT)[g_0 \exp(-\epsilon_0/kT) + g_1 \exp(-\epsilon_1/kT) + \cdots]$$

$$= g_{e_0} \cdot \exp(-e/kT) \cdot f_{\text{vib}}. \tag{3.39}$$

It follows that

$$\ln f'' = \ln g_{e_0} - \frac{e}{kT} + \ln f_{\text{vib}} \tag{3.40}$$

and

$$\frac{\partial \ln f''}{\partial T} = \frac{e}{kT^2} + \frac{\partial \ln f_{\text{vib}}}{\partial T}, \tag{3.41}$$

so that expressions for the Helmholtz function and energy of the crystal become

$$A'' = -NkT \ln f'' = -NkT \ln g_{e_0} + Ne - NkT \ln f_{\text{vib}}$$

$$= -NkT \ln g_{e_0} + Ne + \tfrac{3}{2}Nk\theta + 3NkT \ln[1 - \exp(-\theta/T)], \tag{3.42}$$

and

$$E'' = NkT^2 \left(\frac{\partial \ln f''}{\partial T}\right)_{V/N} = Ne + NkT^2 \left(\frac{\partial \ln f_{\text{vib}}}{\partial T}\right)_{V/N}$$

$$= Ne + \tfrac{3}{2}Nk\theta + \frac{3Nk\theta}{\exp(\theta/T) - 1}. \tag{3.43}$$

The quantity $Ne + \tfrac{3}{2}Nk\theta$ is the negative of the energy of sublimation of the crystal at absolute zero, so that we may write

$$\Delta E^0_{\text{subm}} = -Ne - \tfrac{3}{2}Nk\theta. \tag{3.44}$$

It is not perhaps obvious at first sight why the second term figures in the expression for the energy of sublimation. The point is that the vibrational energy of any particle in a condensed phase is non-zero because of the close proximity of its neighbours. The atoms when separated no longer come under such influence and so no longer possess such vibrational energy, so that in the sublimation process at absolute zero the vibrational ground state energy is 'available' to be used as part of the necessary sublimation energy.

We shall wish, in Chapter 9 to calculate the equilibrium vapour pressure of a monatomic solid by equating its chemical potential to that of the gas. It is presumably clear from what we have discussed above that the expression which must be used for the chemical potential of the crystal is that which derives from A'', i.e.

$$\mu = \left(\frac{\partial A''}{\partial N}\right)_{T,V}$$

$$= -kT \ln g_{e_0} + e + \tfrac{3}{2} k\theta + 3kT \ln [1 - \exp(-\theta/T)]$$

$$= -kT \ln g_{e_0} - \frac{\Delta E_{subm}^0}{N} + 3kT \ln [1 - \exp(-\theta/T)]. \qquad (3.45)$$

We lastly point out that whether we choose to set our energy zero at the hypothetical state in which the atoms in the crystal have zero vibrational energy, or in the state in which all atoms are in their lowest vibrational energy level, and whether we take the potential energy Ne into account or not is quite immaterial in the formulation of expressions for the entropy of the system (because neither the energies $3Nk\theta/2$ nor Ne are available for distribution among the atoms) and in formulating expressions for the heat capacity (because these energies are independent of temperature).

3.6   The Boltzmann Distribution Law

In Chapter 2 it was shown that the most probable distribution of N independent localised particles over accessible energy states is such that the number of particles characterised by an energy ϵ_i is given by the equation

$$n_i = \frac{N}{f} g_i \exp(-\beta \epsilon_i).$$

In view of the identification of β with $1/kT$, this equation may of course be written as

$$n_i = \frac{N}{f} g_i \exp(-\epsilon_i/kT).$$

In the next chapter it will be shown that the distribution of independent *non*-localised particles is described by identical equations.

We have already shown that these equations, which are usually known as the Boltzmann distribution law, play a vital rôle in the construction of the whole body of relationships so far obtained, but this apart, they lead to other useful results and it is to these that attention is now drawn.

The first is that the ratio of the number of particles characterised by energy ϵ_i to that characterised by energy ϵ_j is given by the equation

$$\frac{n_i}{n_j} = \frac{g_i}{g_j} \exp\left[(\epsilon_j - \epsilon_i)/kT\right] \tag{3.46}$$

but that the number of particles (m_i) in *any one quantum-mechanical state* characterised by energy ϵ_i to the number (m_j) in *any one quantum-mechanical state* characterised by energy ϵ_j is given by the equation

$$\frac{m_i}{m_j} = \exp\left[(\epsilon_j - \epsilon_i)/kT\right]. \tag{3.47}$$

It follows that if ϵ_j is greater than ϵ_i, m_i is greater than m_j, *but that n_i is not necessarily greater than n_j*, because of the presence in (3.46) of the degeneracies g_i and g_j. It is in fact usual to find that the higher the energy level the greater is its degeneracy, and this usually produces a 'peak' in the distribution 'curve' such as those which will be seen in Figure 3.4.

It is instructive to calculate the population numbers for various energy levels for assemblies of localised simple harmonic oscillators to demonstrate the effect of first degeneracy, and second of temperature on the distribution.

We first consider an assembly of N localised one-dimensional simple harmonic oscillators, the possible energy levels of which are $(i + \frac{1}{2})h\nu$, i.e. $(i + \frac{1}{2})k\theta$, and recall that these levels are non-degenerate, and calculate the population numbers n_i for $\epsilon_0, \epsilon_1, \epsilon_2, \ldots$ for the temperature $T = \theta$. It is easily shown that we have the following relations:

$$n_0/n_1 = n_1/n_2 = n_2/n_3 = \cdots = e = 2.7183$$

$$N = \sum_i n_i = n_0 + n_0/e + n_0/e^2 \cdots$$

$$= n_0/(1 - e^{-1})$$

so that

$$n_0/N = 1 - e^{-1}, \qquad n_1/N = n_0/Ne \cdots .$$

The result of these calculations is given in Table 3.3 and shown graphically in Figure 3.3(i).

For comparison we consider the distribution of N localised isotropic three-dimensional simple harmonic oscillators at the same temperature,

Table 3.3

ϵ_i	n_i/N at	
	$T = \theta$	$T = 2\theta$
ϵ_0	0.632	0.394
ϵ_1	0.232	0.239
ϵ_2	0.0855	0.145
ϵ_3	0.0315	0.0880
ϵ_4	0.0116	0.0533
ϵ_5	0.0043	0.0324
ϵ_6	0.0016	0.0196
ϵ_7	0.0006	0.0119
ϵ_8	0.0002	0.0072
ϵ_9	0.0001	0.0044
ϵ_{10}	0.0000	0.0027

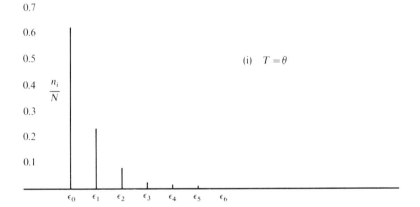

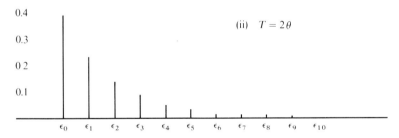

Figure 3.3   The distribution of localised one-dimensional oscillators among non-degenerate energy levels.

recalling that the possible energy levels are $(i + \frac{3}{2})h\upsilon$, i.e. $(i + \frac{3}{2})k\theta$, and that the degeneracy of each level is given by the equation

We here have the general relation

$$n_i/n_{i+1} = (g_i/g_{i+1})\exp(\theta/T) = g_ie/g_{i+1},$$

and it may easily be shown that

$$n_0/n_1 = e/3, \qquad n_1/n_2 = e/2, \qquad n_2/n_3 = 6e/10, \ldots$$

and that $n_0/N = (1 - e^{-1})^3$, from which formulae arise the values for n_i/N given in Table 3.4, the results being shown graphically in Figure 3.4(i).

Table 3.4

		n_i/N at	
ϵ_i	g_i	$T = 0$	$T = 20$
ϵ_0	1	0.2524	0.0608
ϵ_1	3	0.2786	0.1107
ϵ_2	6	0.2049	0.1343
ϵ_3	10	0.1256	0.1358
ϵ_4	15	0.0693	0.1236
ϵ_5	21	0.0357	0.1050
ϵ_6	28	0.0175	0.0850
ϵ_7	36	0.0083	0.0663
ϵ_8	45	0.0038	0.0503
ϵ_9	55	0.0017	0.0373
ϵ_{10}	66	0.0007	0.0272

Comparison of Figures 3.3(i) and 3.4(i) shows two features of note. First the very much greater 'spread' in the second case, and second, the appearance of the 'peak' in the distribution 'curve'. When we come to consider the distribution of non-localised particles (the energy levels for which are necessarily highly degenerate) we shall find this behaviour even more clearly pronounced. We lastly point out that since the behaviour of an assembly of localised three-dimensional oscillators represents quite closely the behaviour of a real monatomic solid, the distribution of atoms within the crystal at $T = 0$ would be expected to be very near that distribution shown in Figure 3.4(i).

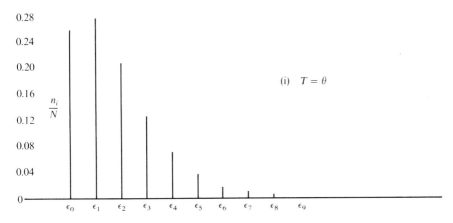

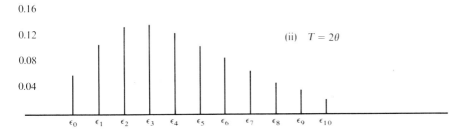

Figure 3.4   The distribution of localised three-dimensional oscillators among degenerate energy levels. $g_i = \frac{1}{2}(i + 1)(i + 2)$

We now raise the interesting question as to the dependence of the distribution numbers on temperature. From the distribution law

$$\ln (n_i/N) = \ln g_i - \epsilon_i/kT - \ln f.$$

The degeneracies are not dependent on temperature, so that

$$\frac{\partial \ln (n_i/N)}{\partial T} = \frac{\epsilon_i}{kT^2} - \frac{\partial \ln f}{\partial T}$$

$$= \epsilon_i/kT^2 - E/NkT^2 \qquad \text{(from 2.29)}$$

$$= \frac{1}{kT^2}(\epsilon_i - \bar{\epsilon})$$

where $\bar{\epsilon}$ is the *mean* energy of the oscillator.

We see that an increase in temperature will *decrease* the population of those energy levels less than the mean, and *increase* the population of those levels greater than the mean. This is borne out by the last columns in Tables 3.3 and 3.4 which show the distribution numbers for the temperature $T = 2\theta$, and by Figures 3.3(ii) and 3.4(ii). The most important feature which emerges from these diagrams is the increase in 'spread' resulting from rise in temperature. It is of course this increase which is represented by the increase in entropy of the assembly when the temperature is raised.

We may calculate the entropy of an assembly of oscillators from the distribution numbers given in Tables 3.3 and 3.4 and it is perhaps instructive to do so to remind ourselves that although we have produced other expressions for the entropy of the assembly (such as (2.30)) such expressions are merely *more convenient* ways of expressing the relationship

$$S = k \ln W_{total} = k \ln W_{max}.$$

For an assembly of N localised one-dimensional oscillators, the energy levels of which are non-degenerate,

$$\frac{S}{k} = \ln \frac{N!}{n_0! \, n_1! \, n_2! \cdots n_i!}$$

so that the entropy at $T = \theta$ may be calculated (using the Stirling formula) from the appropriate figures in Table 3.3. The result for $N = 10^6$ (neglecting in the denominator terms $n_{10}!$ and beyond) is

$$S = 1.049 \times 10^6 k. \tag{3.48}$$

We may do the same for an assembly of N localised three-dimensional oscillators using the appropriate figures in Table 3.4 and the equation

$$\frac{S}{k} = \ln \frac{N! \prod_i g_i^{n_i}}{\prod_i n_i!}.$$

The result for $N = 10^6$ and $T = \theta$ is

$$S = 3.140 \times 10^6 k. \tag{3.49}$$

The reader should be clear why this result is almost exactly three times that obtained in (3.48). (The fact that one value is *not exactly* three times the other is due to the neglect of terms $n_{10}!$ and beyond.)

There remains one further use of the distribution law to which attention should be drawn, and that is the determination of the number of particles occupying energy levels greater than any chosen one. Denoting the total number of particles occupying energy level i and levels higher than i by the symbol $N \geqslant \epsilon_i$, we require an expression for the fraction $N \geqslant \epsilon_i / N$. This is obtained easily in the case of an assembly of localised one-dimensional oscillators.

Since

$$n_i/N = \frac{\exp\left[-(i + \tfrac{1}{2})\theta/T\right]}{f_{\text{odo}}}$$

$$\frac{N \geqslant \epsilon_i}{N} = \frac{\exp\left[-(i + \tfrac{1}{2})\theta/T\right] + \exp\left[-(i + \tfrac{3}{2})\theta/T\right] + \exp\left[-(i + \tfrac{5}{2})\theta/T\right] + \cdots}{f_{\text{odo}}}$$

$$= \frac{\exp\left[-(i + \tfrac{1}{2})\theta/T\right][1 + \exp(-\theta/T) + \exp(-2\theta/T) + \cdots]}{\exp(-\theta/2T)[1 + \exp(-\theta/T) + \exp(-2\theta/T) + \cdots]}$$

$$= \exp(-i\theta/T) = \exp(-ih\nu/kT) = \exp\left[-(\epsilon_i - \epsilon_0)/kT\right]$$

$$= \exp(-E/RT) \qquad (3.50)$$

where we have denoted $L(\epsilon_i - \epsilon_0)$ by E, so that E is the energy of one mole of oscillators at energy level ϵ_i in excess of that in their ground state. Equation (3.50) has important applications in such problems as the rate of evaporation of atoms from a solid surface.

The corresponding expression for an assembly of three-dimensional oscillators is less simple. We have the following relations:

$$\epsilon_i = (i + \tfrac{3}{2})h\nu = (i + \tfrac{3}{2})k\theta,$$

$$\frac{n_i}{N} = \frac{g_i \exp\left[-(i + \tfrac{3}{2})\theta/T\right]}{f_{\text{tdo}}},$$

$$f_{\text{tdo}} = \left[\frac{\exp(-\theta/2T)}{1 - \exp(-\theta/T)}\right]^3,$$

and

$$g_i = \tfrac{1}{2}(i + 1)(i + 2),$$

from which it follows that

$$\frac{N \geqslant \epsilon_i}{N} = \frac{\exp -(i + \tfrac{3}{2})\theta/T[g_i + g_{i+1}\exp(-\theta/T) + g_{i+2}\exp(-2\theta/T) + \cdots]}{\dfrac{\exp(-3\theta/2T)}{[1 - \exp(-\theta/T)]^3}}$$

$$= \exp(-i\theta/T)(1 - x)^3(g_i + g_{i+1}x + g_{i+2}x^2 + \cdots)$$

where $x = \exp(-\theta/T)$. Hence

$$\frac{N \geqslant \epsilon_i}{N} = \exp(-i\theta/T)[\tfrac{1}{2}(i + 1)(i + 2) - (i^2 + 2i)x + \tfrac{1}{2}(i^2 + i)x^2]$$

with all higher terms vanishing. We have therefore

$$\frac{N \geqslant \epsilon_i}{N} = \exp\left(-i\theta/T\right)\left[\tfrac{1}{2}(i+1)(i+2) - (i^2 + 2i)\exp\left(-\theta/T\right)\right.$$

$$\left. +\tfrac{1}{2}(i^2 + i)\exp\left(-2\theta/T\right)\right]. \tag{3.51}$$

EXERCISES

3.1 Obtain the value for the entropy of an assembly of 10^6 localised one-dimensional oscillators at $T = \theta$ given in (3.48) and compare it with the value obtained using equations
(2.30) and (3.7). Why the very small discrepancy?

3.2 Similarly obtain the value for the entropy of an assembly of 10^6 localised three-dimensional
oscillators at $T = \theta$ given in (3.49) and compare this with that obtained using equations
(2.30) and (3.11).

CHAPTER 4

An Assembly of Non-Localised Particles

4.1 Introduction

The aim of this chapter, and of the next, is to obtain expressions for the thermodynamic properties of a perfect gas. The model we choose to represent the system is an assembly of N identical, independent, non-localised particles. We shall proceed in very much the same way as we did in Chapters 2 and 3 when studying the assembly of localised particles: (i) obtain an expression for the total number of complexions accessible to the system and for the number corresponding to the most probable configuration, the energy and volume of the system being considered fixed, (ii) find, as we did before, that a quantity known as the partition function of the particle emerges quite naturally from our equations, (iii) express the properties of the assembly in terms of the partition function, and (iv), in the next chapter, show how to express the partition function in terms of the characteristics of the molecule, its mass, its moments of inertia, the forces governing intra-molecular vibrations and so on. It may be remarked that although operation (i) proves to be rather less straight-forward than did the corresponding operation in Chapter 2, operation (iv) proves to be simpler (and more accurate) than the corresponding operation in the case of the crystal (the Einstein and Debye treatments), and the theoretical predictions for the values of the properties of a gaseous system are precisely in accord with the experimental values. It is certainly true that such predictions are easy only for relatively simple molecules, but any difficulty found in predicting results for more complicated molecules lies not at all in the *statistical–mechanical* field, but in unravelling some of the experimental (spectroscopic) data leading to values for the molecular characteristics.

4.2   The Number of Complexions
Accessible to the Assembly

Much of the success of the method to be followed in this chapter stems from the fact that a perfect gas is *by definition* the assembly of independent particles required by our treatment. The word 'independent' signifies of course just what it did in Chapter 2, that the particles may exchange energy as the result of intermolecular collisions, but that at any one instant the quantum state of any one is independent of that of its neighbours.

The model we set up differs fundamentally however from that set up in Chapter 2 in two respects. The first is that the particles are *non-localised*, and it is this that (at first sight) presents difficulties in obtaining expressions for W_{total} and W_{max}. In a crystal each particle occupies a particular *site*: accordingly the interchange of two like particles in different energy states leads to two recognisably different complexions for the crystal as a whole, for the particle sites are distinguishable even if the particles are not, whereas the mention of a particular site in relation to the molecules of a gas is meaningless, the entire volume of the system being accessible to them all. The second (and it is this which relieves the situation and makes progress possible) is that *all energy levels accessible to non-localised particles are necessarily highly degenerate* (at least, at all temperatures except those very close indeed to absolute zero.) This condition stems essentially from the fact that non-localised particles necessarily possess translational energy (otherwise they would be localised), and from the fact that all translational energy levels are necessarily highly degenerate. This latter fact is easily demonstrated.

The velocity of C of an independent non-localised particle may be resolved into three vectors C_x, C_y and C_z parallel to three mutually perpendicular axes X, Y and Z so that

$$C^2 = C_x^2 + C_y^2 + C_z^2.$$

The translational energy $\frac{1}{2}mC^2$ may similarly be resolved to give

$$\tfrac{1}{2}mC^2 = \tfrac{1}{2}mC_x^2 + \tfrac{1}{2}mC_y^2 + \tfrac{1}{2}mC_z^2$$

i.e.

$$\epsilon_t = \epsilon_x + \epsilon_y + \epsilon_z. \tag{4.1}$$

The physical properties of a gas are unaffected by the *shape* of their container, so we are entitled to consider the behaviour of a molecule in a container shaped for our convenience, so we will consider an assembly of hydrogen molecules in a cubic container of sides x, y and x ($x = y = z = 1$ cm) the sides being parallel to the axes X, Y and Z, and suppose the temperature of the assembly to be 300 K. The component of the translational energy of any one molecule in direction X is given by the equation

$$\epsilon_x = \tfrac{1}{2}mC_x^2 = \frac{1}{2m}(mC_x)^2. \tag{4.2}$$

Quantum mechanics relates the momentum of a particle to an associated wavelength λ by the de Broglie equation which states that the product of the momentum and the wavelength equals Planck's constant h, so that we may write

$$\epsilon_x = \frac{1}{2m}\left(\frac{h}{\lambda}\right)^2. \tag{4.3}$$

When the particle is in a container of rigid sides, the associated wave motion must be such that its amplitude is zero at the sides, so that the distance between the sides must be an integral number of half wavelengths, i.e.

$$x = \frac{n\lambda}{2}, \tag{4.4}$$

where n can assume any integral value from unity to infinity. It follows that

$$\epsilon_x = \frac{h^2 n^2}{8mx^2}. \tag{4.5}$$

Similar equations may be obtained for ϵ_y and ϵ_z. The translational energy of a single molecule can therefore have any value given by the equation

$$\epsilon_t = \epsilon_x + \epsilon_y + \epsilon_z$$

$$= \frac{h^2}{8m}\left[\frac{n_x^2}{x^2} + \frac{n_y^2}{y^2} + \frac{n_z^2}{z^2}\right] \tag{4.6}$$

the three translational quantum numbers n_x, n_y and n_z being permitted to assume any integral value from unity to infinity.

We shall see later that the total translational energy of an assembly of N molecules at temperature T is $\tfrac{3}{2}NkT$, where k is Boltzmann's constant, so that the mean value for the translational energy for a single molecule is $\tfrac{3}{2}kT$, and so is, at 300 K,

$$\tfrac{3}{2} \times 1.38 \times 10^{-23} \times 300 \text{ J}, \quad \text{i.e. } 6.2 \times 10^{-21} \text{ J}.$$

The mass of the hydrogen molecule is 3.2×10^{-27} kg, h is approximately 6.6×10^{-34} Js, and we have chosen a container of sides one centimetre, so that the average value for $[n_x^2 + n_y^2 + n_z^2]$ for hydrogen molecules at

300 K in a cubic container of sides one centimetre is

$$\frac{8 \times 3.2 \times 6.2}{6.6 \times 6.6} \times 10^{20} \times 10^{-4}, \quad \text{i.e. } 3.6 \times 10^{16}.$$

This means that a hydrogen molecule in such a container at such a temperature possesses the mean value for translational energy if it selects any integral numbers n_x, n_y and n_z satisfying the equation

$$n_x^2 + n_y^2 + n_z^2 = 3.6 \times 10^{16}. \tag{4.7}$$

Each particular set of numbers n_x, n_y, n_z defines a particular translational state. The number of ways in which three numbers n_x, n_y and n_z can be chosen so that equation (4.7) is satisfied is the value for the degeneracy of the mean translational energy level.

We set out to demonstrate that any energy level accessible to a molecule in a gas is highly degenerate. We may consider the proposition proved.

We are now in a position to set up formally the model to be used in our calculations. We consider an assembly of N non-localised (and hence indistinguishable) particles in a container and suppose the total energy and volume fixed. We further suppose that at some instant n_0 particles occupy energy level ϵ_0, n_1 occupy level $\epsilon_1, \ldots, n_i$ occupy level $\epsilon_i, \ldots$, the degeneracies of each level being $g_0, g_1, \ldots, g_i, \ldots$, no restriction being placed on the energy levels (in the sense that each may be regarded as the sum of possible translational, rotational, vibrational energies ... and so on). We now set out to calculate the number of complexions accessible to the system, but before doing so must consider a possible complication forced upon us by quantum mechanics.

Quantum mechanics shows that any given assembly of *non-localised* particles obeys *one* of two types of statistics, one developed by Fermi and Dirac, and the other by Bose and Einstein. To understand the fundamental difference between them requires a greater knowledge of quantum mechanics than is assumed in this book, but a distinction which is of some practical utility as far as we are concerned is that Fermi–Dirac statistics apply to those systems in which any particular quantum state can be occupied only by one particle at a time (this restriction is clearly linked to the Pauli Exclusion Principle), whilst Bose–Einstein statistics apply to those systems in which no restriction governs the number of particles occupying any particular quantum state.

Let us first assume that the assembly with which we are concerned is governed by Fermi–Dirac statistics, which means that (considering the occupancy of energy level ϵ_i) each of the g_i quantum states available to the particles can be occupied by only one particle or none. The number of ways of assigning n_i indistinguishable objects to g_i distinguishable states so that each state is occupied only by one particle or by none is

$$\frac{g_i!}{n_i!\,(g_i - n_i)!}.$$

The number of ways in which the whole assembly may be reached is the product of these terms for all energy levels, i.e.

$$W_{FD} = \prod_i \frac{g_i!}{n_i!\,(g_i - n_i)!}$$

$$= \prod_i \frac{g_i(g_i - 1)(g_i - 2)\cdots(g_i - n_i + 1)}{n_i!}. \tag{4.8}$$

Now as we demonstrated above the degeneracy of any energy level accessible to any molecule in a gas (except at *very* low temperatures) is a very large number, and it may be shown that for all levels ϵ_i, g_i is very much greater than n_i, the number of molecules likely to be found therein. In these circumstances each term in the numerator of (4.8) approximates to g_i, and since there are n_i such terms

$$W_{FD} \simeq \prod_i \frac{g_i^{n_i}}{n_i!}. \tag{4.9}$$

Let us now suppose that the assembly is governed by Bose–Einstein statistics. Here we are concerned with the number of ways of assigning n_i indistinguishable objects to g_i distinguishable states, there being no restriction as to the number of objects in any one. The answer is

$$\frac{(g_i + n_i - 1)!}{(g_i - 1)!\,n_i!},$$

so that the number of complexions for the configuration $n_0, n_1, \ldots, n_i, \ldots$ is given by the expression

$$W_{BE} \simeq \prod_i \frac{(g_i + n_i - 1)!}{(g_i - 1)!\,n_i!}$$

$$= \prod_i \frac{(g_i + n_i - 1)(g_i + n_i - 2)\cdots(g_i + 1)g_i}{n_i!}$$

$$\simeq \prod_i \frac{g_i^{n_i}}{n_i!}. \tag{4.10}$$

These results show that as long as each g_i greatly exceeds the corresponding population number n_i, the W value for any one configuration $n_0, n_1, \ldots,$ $n_i, \ldots$ is for all practical purposes the same whichever statistics govern the system, and that this is entirely reasonable can be seen, mathematics apart, it is obviously unnecessary to inquire whether a particular quantum state is *permitted* to be occupied by one or by any number of particles, because if the number of states greatly exceeds the number of particles available to fill them, chance will ensure that each state is occupied by one particle at the most.[1]

It appears therefore that the total number of complexions accessible to the assembly (taking all possible distributions into account) is given by the equation

$$W_{\text{total}} = \prod_i \frac{g_i^{n_i}}{n_i!} + \prod_i \frac{g_i^{n_i'}}{n_i'!} + \prod_i \frac{g_i^{n_i''}}{n_i''!} + \cdots \qquad (4.11)$$

the first term representing the number of complexions leading to the configuration governed by the set of population numbers n_i, the second term the number leading to the configuration governed by the set n_i', and so on. This equation is of course analogous to (2.4) in Chapter 2, and just as we did then we proceed by picking out the most probable configuration and writing

$$\ln W_{\text{total}} \simeq \ln W_{\text{max}} = \ln\left[\prod_i \frac{g_i^{n_i^*}}{n_i^*!}\right] \qquad (4.12)$$

where n_i^* is the set of population numbers leading to the most probable configuration, the set n_i^* being determined by the equation

$$\delta \ln W = \left(\frac{\partial \ln W}{\partial n_0}\right)_{n_j} \delta n_0 + \left(\frac{\partial \ln W}{\partial n_1}\right)_{n_j} \delta n_1 + \cdots + \left(\frac{\partial \ln W}{\partial n_i}\right)_{n_j} \delta n_i + \cdots = 0$$

$$(4.13)$$

the increments $\delta n_0, \delta n_1, \ldots, \delta n_i, \ldots$ being of course subject to the conditions

$$\delta E = \epsilon_0 \delta n_0 + \epsilon_1 \delta n_1 + \cdots + \epsilon_i \delta n_i + \cdots = 0 \qquad (4.14)$$

$$\delta N = \delta n_0 + \delta n_1 + \cdots + \delta n_i + \cdots = 0. \qquad (4.15)$$

Proceeding exactly as we did in Chapter 2, multiplying (4.15) by α, (4.14) by $-\beta$ and adding the products to (4.13) we find that the values $n_0^*, n_1^*, \ldots,$

---

[1] None the less it must be realised that the assumption that $g_i \gg n_i$ cannot be made if the temperature of the assembly is *very* close to absolute zero, because at very low temperatures most of the particles occupy the lowest energy levels the degeneracies of which are not high, and therefore the formulae we shall produce cannot be extrapolated to predict the properties of a system at absolute zero, as indeed we shall see from their form. Neither can this assumption be made if we are considering the distribution of electrons in a metal. Such a system is governed strictly by Fermi–Dirac statistics.

$n_i^*, \ldots$ are those values of $n_0, n_1, \ldots, n_i, \ldots$ given by the equations

$$\frac{\partial \ln W}{\partial n_0} + \alpha - \beta \epsilon_0 = 0$$

$$\frac{\partial \ln W}{\partial n_1} + \alpha - \beta \epsilon_1 = 0 \qquad (4.16)$$

$$\vdots$$

$$\frac{\partial \ln W}{\partial n_i} + \alpha - \beta \epsilon_i = 0$$

which are of course the same as equations (2.12). From (4.12)

$$\ln W = \sum_i n_i \ln g_i - \sum_i \ln n_i!$$

and therefore, making use of the Stirling approximation formula,

$$\ln W = \sum_i n_i \ln g_i - \sum_i n_i \ln n_i + \sum_i n_i \qquad (4.17)$$

and

$$\frac{\partial \ln W}{\partial n_i} = \ln \frac{g_i}{n_i}. \qquad (4.18)$$

Substituting into equations (4.16) we see that the population numbers leading to the most probable distribution are those given by the equations

$$\ln \frac{n_0^*}{g_0} = \alpha - \beta \epsilon_0 \qquad (4.19)$$

$$\ln \frac{n_1^*}{g_1} = \alpha - \beta \epsilon_1 \qquad (4.20)$$

$$\ln \frac{n_i^*}{g_i} = \alpha - \beta \epsilon_i \qquad (4.21)$$

so that for each energy level

$$n_i^* = g_i \exp(\alpha) \exp(-\beta \epsilon_i). \qquad (4.22)$$

Since

$$N = \sum_i n_i^* = \exp(\alpha) \sum_i g_i \exp(-\beta \epsilon_i)$$

we may eliminate $\exp(\alpha)$ from (4.22) and write

$$\begin{aligned}
\frac{n_i^*}{N} &= \frac{g_i \exp(-\beta\epsilon_i)}{\sum_i g_i \exp(-\beta\epsilon_i)} \\
&= \frac{g_i \exp(-\beta\epsilon_i)}{f}
\end{aligned} \tag{4.23}$$

where we define f by the equation

$$f = \sum_i g_i \exp(-\beta\epsilon_i) \tag{4.24}$$

recalling that the symbol $\sum_i$ denotes summation over all energy *levels*. Here again the quantity f is known as the partition function of the particle. It will be noticed that the expression for the Boltzmann Distribution law (4.23) and that for the partition function (4.24) are precisely the same as those expressions for the localised particles. Attention is drawn however to the difference in the expressions for W_{max}, i.e. the difference between equations (4.12) and (2.5). The *absence* of the term $N!$ in (4.12) is due to the essential indistinguishability of non-localised particles.

4.3   The Thermodynamic Properties of an Assembly of Non-Localised Particles

We are now in a position to obtain expressions for the properties of an isolated assembly of N independent non-localised particles in terms of the partition function. Denoting the population numbers leading to the most probable configuration by the symbols $n_0, n_1, \ldots, n_i, \ldots$ we have the following relations:

$$N = \sum_i n_i \tag{4.25}$$

$$E = \sum_i n_i \epsilon_i \tag{4.26}$$

$$n_i = \frac{N}{f} g_i \exp(-\beta\epsilon_i) \tag{4.27}$$

$$f = \sum_i g_i \exp(-\beta\epsilon_i) \tag{4.28}$$

and may immediately write

$$\frac{S}{k} = \ln W_{total} \simeq \ln W_{max} \simeq \sum_i \left(\ln \frac{g_i^{n_i}}{n_i!} \right).$$

We again make use of the Stirling approximation formula and obtain the equation

$$\frac{S}{k} = \sum_i \left(n_i \ln \frac{g_i}{n_i} + n_i \right). \tag{4.29}$$

It follows from (4.27) that

$$\ln \frac{g_i}{n_i} = \ln \frac{f}{N} + \beta \epsilon_i$$

so that

$$\frac{S}{k} = N \ln \frac{f}{N} + \beta E + N. \tag{4.30}$$

We shall leave it to the reader to show that

$$\left(\frac{\partial S}{\partial E} \right)_{V,N} = \beta k$$

so that here again

$$\beta = \frac{1}{kT}, \tag{4.31}$$

and that

$$E = kNT^2 \left(\frac{\partial \ln f}{\partial T} \right)_V. \tag{4.32}$$

Equation (4.30) may now be expressed in the rather more convenient form

$$S = kN \ln \frac{f}{N} + kNT \left(\frac{\partial \ln f}{\partial T} \right)_V + kN. \tag{4.33}$$

It follows from (4.32) and (4.33) that

$$A = E - TS$$

$$= -kNT \ln \frac{f}{N} - kNT. \tag{4.34}$$

Formal expressions for all other properties of the assembly could now be obtained through the appropriate classical thermodynamic relationships, but these are best left until we are in a position to express f in terms of the physical characteristics of the molecule.

Equations (4.32), (4.33) and (4.34) should be compared with the corresponding equations for the energy, entropy and Helmholtz function for an assembly of localised particles, equations (2.29), (2.30) and (2.31). It will be seen that the expressions for the energies of the assemblies are the same, but that that for the entropy of the assembly of non-localised particles contains the terms $-kN \ln N + kN$ (i.e. $-k \ln N!$) which are absent from (2.30), whilst that for the Helmholtz function for the assembly of non-localised particles contains the terms $+kNT \ln N - kNT$ (i.e. $+kT \ln N!$) which are absent from (2.31). These terms stem from the inherent indistinguishability of non-localised particles.

4.4   Factorisation of the Partition Function

The quantity ϵ_i which appears in the expression

$$f = \sum_i g_i \exp(-\beta \epsilon_i) \qquad (4.28)$$

represents any one of the *total* energy levels accessible to the particle, and must therefore include all of the energies associated with each of its various modes of behaviour. For simplicity, let us consider the possible modes of behaviour of a single non-localised diatomic molecule existing with others of like kind in a container of fixed dimensions. The molecule is continuously moving from one part of the container to another, so possessing at any one moment any one of a range of possible translational energies ϵ_t, it may rotate about its centre of mass so possessing any one of a range of rotational energies ϵ_r, it is subject to continuous changes in the distance separating the nuclei of its atoms so possessing any one of a range of possible vibrational energies ϵ_v, it may possess any one of a range of possible electronic energies ϵ_e related to the forces between its electrons and nuclei, and, in principle at least, we must suppose that each atomic nucleus can exist in any one of a number of nuclear energy levels ϵ_{n_1} (related to atom 1) and ϵ_{n_2} (related to atom 2).

If we assume that the energies associated with any one mode of behaviour are independent of those associated with all others we may describe any typical energy level ϵ_i by the equation

$$\epsilon_i = \epsilon_t + \epsilon_r + \epsilon_v + \epsilon_e + \epsilon_{n_1} + \epsilon_{n_2} \qquad (4.35)$$

(it being recognised that the same value ϵ_i may be reached by different combinations of the individual contributions), and g_i, the degeneracy of that level by the equation

$$g_i = g_t \cdot g_r \cdot g_v \cdot g_e \cdot g_{n_1} \cdot g_{n_2} \qquad (4.36)$$

where g_t is the number of translational states associated with translational energy ϵ_t, g_r the number of distinguishable rotational states associated with rotational energy $\epsilon_r, \ldots$ and so on. In other words $g_t, g_r, \ldots$ are the respective degeneracies of the translational energy level ϵ_t, and of the rotational energy level $\epsilon_r, \ldots$.

It follows that (4.28) may be expanded into the form

$$f = \sum_i g_t \cdot g_r \cdot g_v \cdot g_e \cdot g_{n_1} \cdot g_{n_2} \exp\left[-\beta(\epsilon_t + \epsilon_r + \epsilon_v + \epsilon_e + \epsilon_{n_1} + \epsilon_{n_2})\right]$$

$$= \left[\sum_t g_t \exp\left(-\beta\epsilon_t\right)\right]\left[\sum_r g_r \exp\left(-\beta\epsilon_r\right)\right]\left[\sum_v g_v \exp\left(-\beta\epsilon_v\right)\right]$$

$$\times \left[\sum_e g_e \exp\left(-\beta\epsilon_e\right)\right]\left[\sum_{n_1} g_{n_1} \exp\left(-\beta\epsilon_{n_1}\right)\right]\left[\sum_{n_2} g_{n_2} \exp\left(-\beta\epsilon_{n_2}\right)\right] \quad (4.37)$$

and we are obviously justified in regarding the sum in the first bracket as the translational partition function for the molecule, the sum in the second as being the rotational partition function . . . and so on, and so write

$$f = f_t \cdot f_r \cdot f_v \cdot f_e \cdot f_{n_1} \cdot f_{n_2}. \quad (4.38)$$

The advantages that accrue from this are considerable. The first is that each of the quantities $f_t, f_r, \ldots$ may be evaluated (or expressed in terms of the appropriate physical characteristic of the molecule) without reference to the remainder. The second results from the fact that all expressions for the properties of the assembly are in terms of $\ln f$ rather than of f itself, and since from (4.38)

$$\ln f = \ln f_t + \ln f_r + \ln f_v + \ln f_e + \ln f_{n_1} + \ln f_{n_2} \quad (4.39)$$

we can easily obtain expressions for the separate contributions to any thermodynamic property from each mode of behaviour. Thus from (4.32) we see that

$$E = kNT^2\left(\frac{\partial \ln f_t}{\partial T}\right)_V + kNT^2\frac{d \ln f_r}{dT} + kNT^2\frac{d \ln f_v}{dT}$$

$$+ kNT^2\frac{d \ln f_e}{dT} + kNT^2\frac{d \ln f_{n_1}}{dT} + kNT^2\frac{d \ln f_{n_2}}{dT} \quad (4.40)$$

the first term being the translational contribution, the second the rotational contribution . . . and from (4.33)

$$S = \left[kN \ln \frac{f_t}{N} + kNT\left(\frac{\partial \ln f_t}{\partial T}\right)_V + kN\right]$$

$$+ \left[kN \ln f_r + kNT\frac{d \ln f_r}{dT}\right] + \left[kN \ln f_v + kNT\frac{d \ln f_v}{dT}\right]$$

$$+\left[kN\ln f_e + kNT\frac{d\ln f_e}{dT}\right] + \left[kN\ln f_{n_1} + kNT\frac{d\ln f_{n_1}}{dT}\right]$$

$$+\left[kN\ln f_{n_2} + kNT\frac{d\ln f_{n_2}}{dT}\right] \tag{4.41}$$

the terms in the first bracket being the translational contribution to the entropy of the assembly, the terms in the second the rotational contribution and so on.

These equations call for two remarks. The first is that as we shall find in the next chapter, the value of only the translational partition function depends on the volume to which the particle is confined, so we have correctly written a partial differential only in the first term in (4.40) and in the second term of the first bracket in (4.41). The second is that it will be seen that we have included the terms $-kN\ln N + kN$ in the expression for the translational contribution to the entropy of the assembly. This is appropriate because, as mentioned earlier, both terms arise because of the indistinguishability of non-localised particles and so from their translational motion.

For the third advantage which accrues from (4.38) we return to the Distribution Law,

$$n_i = \frac{N}{f}g_i \exp(-\beta\epsilon_i).$$

Consider an assembly of particles each of which we suppose for simplicity to be capable of only two modes of activity a and b. Then

$$\epsilon_i = \epsilon_a + \epsilon_b, \qquad g_i = g_a \cdot g_b \quad \text{and} \quad f = f_a \cdot f_b.$$

The number of particles possessing both a particular value of ϵ_a say ϵ_{a*} and a particular value of ϵ_b say ϵ_{b*} is given by the expression

$$n_{a*,b*} = \frac{Ng_{a*}g_{b*}\exp[-\beta(\epsilon_{a*} + \epsilon_{b*})]}{f_a f_b}$$

$$= N\left[\frac{g_{a*}\exp(-\beta\epsilon_{a*})}{f_a}\right]\left[\frac{g_{b*}\exp(-\beta\epsilon_{b*})}{f_b}\right], \tag{4.42}$$

and therefore the number of particles possessing energy ϵ_{a*} whilst possessing *any* value for ϵ_b is

$$n_{a*(\text{all b})} = N\left[\frac{g_{a*}\exp(-\beta\epsilon_{a*})}{f_a}\right]\left[\frac{\sum_b g_b \exp(-\beta\epsilon_b)}{f_b}\right]$$

$$= N\frac{g_{a*}\exp(-\beta\epsilon_{a*})}{f_a}. \tag{4.43}$$

This means that the Distribution Law can be applied to each separate mode of behaviour of the particle without reference to the other modes of activity. Thus the number of particles possessing a particular value of translational energy ϵ_{t*} (irrespective of the energies related to other modes) is given by the equation

$$n_{t*} = \frac{N}{f_t} g_{t*} \exp\left(-\beta \epsilon_{t*}\right) \tag{4.44}$$

the number of particles possessing a particular value of rotational energy ϵ_{r*} by the equation

$$n_{r*} = \frac{N}{f_r} g_{r*} \exp\left(-\beta \epsilon_{r*}\right) \tag{4.45}$$

and so on.

Because the advantages which accrue from (4.38) are so great, we must however inquire rather closely whether we were really justified in making the original assumption on which (4.38) was based, the assumption that it is possible to resolve the total energy of a particle into precise components as we did in (4.35).

There is in fact no doubt that the translational energy of a molecule and the nuclear energy levels of the atoms therein are completely independent of each other and of the other modes of activity, so that we can write

$$\epsilon_i = \epsilon_t + \epsilon_{r,v,e} + \epsilon_{n_1} + \epsilon_{n_2}$$

and

$$f = f_t f_{r,v,e} f_{n_1} f_{n_2},$$

where $\epsilon_{r,v,e}$ and $f_{r,v,e}$ are composite terms related to the particle's rotational, vibrational and electronic modes of activity. It is however clear from physical considerations alone that the resolution of $\epsilon_{r,v,e}$ into three independent components

$$\epsilon_{r,v,e} = \epsilon_r + \epsilon_v + \epsilon_e$$

and hence

$$f_{r,v,e} = f_r f_v f_e$$

is not altogether sound. Consider for example a diatomic molecule. Its permissible rotational energy levels are determined (as we shall see in the next chapter) by its moment of inertia, and this depends on the distance between the two atomic nuclei. Molecular vibrations however lead to continual changes in that distance, so that vibrational activity will effect the permissible rotational energy levels. Fortunately, the separation of vibrational energy levels for most simple molecules is so great that the number of molecules occupying excited vibrational states is small, and the error introduced by assuming that the rotational and vibrational activities are truly independent is usually of little importance.

At first sight, the interdependence of *electronic* energy levels and rotational and vibrational activities appears to be even more pronounced because electronic excitation *changes* the vibrational frequencies and bond lengths (and so the moments of inertia) of the molecule, so that each electronic state has its own 'set' of vibrational and rotational energies. It follows that the complete partition function for the molecule is properly given by the expression

$$f = f_t f_n[g_{e_0} \exp\left(-\epsilon_{e_0}/kT\right)f_{r,v}^0 + g_{e_1} \exp\left(-\epsilon_{e_1}/kT\right)f_{r,v}^1 + \cdots] \quad (4.46)$$

where $f_{r,v}^0$ is the 'combined' rotational and vibrational partition function appropriate to the molecule when in its electronic ground state, $f_{r,v}^1$ the corresponding 'combined' partition function appropriate to the molecule when in its first excited electronic state... and so on. Fortunately the separation of electronic energy levels for almost all molecules is so very great that the number of molecules occupying excited electronic states at temperatures which may be reached in the laboratory is completely negligible. This means that all terms except the first in the bracket in (4.46) may be dropped without significant error.

The ultimate justification for the separation of the electronic, rotational and vibrational partition functions is of course experimental and rests on the fact that if the properties of gases at moderate temperatures are determined (as they will be in the next chapter) first on the assumption that (4.38) *is* justified, and then *precisely*, by using actual energy levels deduced from molecular spectroscopy, there proves to be no significant difference between the two sets of results.

EXERCISES

4.1 In the next chapter we shall show quantum-mechanically that the partition function for an independent non-localised particle is proportional to the volume accessible to it. Show that this conclusion may be reached (quantum mechanics apart) from equation (4.34) using the classical thermodynamic fact that the chemical potential for a pure substance is given both by the equation

$$\mu = \left(\frac{\partial A}{\partial N}\right)_{V,T} \qquad \text{(i)}$$

and by the equation

$$\mu = \frac{G}{N} = \frac{A + PV}{N}. \qquad \text{(ii)}$$

CHAPTER 5

The Partition Function for Independent Non-Localised Molecules and the Properties of Gaseous Systems

In this chapter we obtain expressions for the translational, rotational, vibrational and electronic partition functions for a gaseous molecule assuming that these modes of activity are strictly separable, and from these calculate the various contributions to some of the most important thermodynamic properties of perfect gases. In Section 5.11 we show how these values may be converted into corresponding values for real (imperfect) gases, and in Chapter 9 we show how these values may be used to give semi-statistical values for the corresponding properties of liquids and solids.

5.1 The Translational Partition Function

Consider (as we did in Section 4.2) the behaviour of a molecule mass m in a rectangular container volume V of sides x, y and z, so that

$$V = xyz. \tag{5.1}$$

It was shown in that section that the translational energy of the molecule may assume any value ϵ_t given by the equation

$$\epsilon_t = \frac{h^2}{8m}\left[\frac{n_x^2}{x^2} + \frac{n_y^2}{y^2} + \frac{n_z^2}{z^2}\right] \tag{4.6}$$

where n_x, n_y and n_z are the three translational quantum numbers which may assume any integral value from unity to infinity. Each translational energy *level* is of course characterised by the value of the quantity in brackets, the degeneracy of any one level being the number of ways in which that quantity may be reached through different values of n_x, n_y and n_z . There is however no need to identify each energy level and its degeneracy, because it is clear that each combination of particular values of n_x, n_y and n_z defines a particular translational *state*, and we can obtain an expression for the partition function by summing *over all accessible states*, i.e. by summing $\exp\left(-\epsilon_t/kT\right)$ for all values of n_x, n_y and n_z. It is therefore given by the expression

$$f_t = \sum_{n_x,n_y,n_z} \exp\left[-\frac{h^2}{8mkT}\left(\frac{n_x^2}{x^2} + \frac{n_y^2}{y^2} + \frac{n_z^2}{z^2}\right)\right]$$

$$= \sum_{n_x=1}^{n_x=\infty} \exp\left(-An_x^2/x^2\right) \sum_{n_y=1}^{n_y=\infty} \exp\left(-An_y^2/y^2\right) \sum_{n_z=1}^{n_z=\infty} \exp\left(-An_z^2/z^2\right) \qquad (5.2)$$

where

$$A = \frac{h^2}{8mkT}.$$

We denote the first sum by f_x, the second by f_y and the third by f_z, so that

$$f_t = f_x \cdot f_y \cdot f_z. \qquad (5.3)$$

If we consider a typical case, a molecule of H_2 ($m = 3.2 \times 10^{-27}$ kg) contained between walls 10^{-2} metre apart at 300 K we find that A/x^2 is a dimensionless quantity of order 10^{-16}, and this quantity will be smaller for all other gas molecules (because m will be greater) in a larger container. Thus for the case described

$$f_x = \exp\left(-10^{-16}\right) + \exp\left(-4 \cdot 10^{-16}\right) + \exp\left(-9 \cdot 10^{-16}\right) + \cdots$$

and it is evident that each exponential term is only infinitesimally smaller than unity. The value of f_x is clearly the sum of the areas of the shaded rectangles in Figure 5.1, the area of the first representing the term for $n_x = 1$, that of the second the term $n_x = 2$ and so on, and it is clear that the sum of

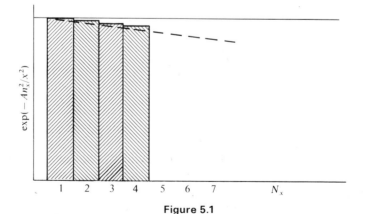

Figure 5.1

these areas is extremely close to that area under the dashed curve[1] which is given by the integral

$$\int_{n_x=0}^{n_x=\infty} \exp\left(-An_x^2/x^2\right) dn_x.$$

All that we have done is to show that because A/x^2 is so much smaller than unity the summation $\sum_{n_x=1}^{n_x=\infty} \exp\left(-An_x^2/x^2\right)$ may be replaced by the integral

$$\int_{n_x=0}^{n_x=\infty} \exp\left(-An_x^2/x^2\right) dn_x.$$

This a standard integral the value of which is $\frac{1}{2}[\pi x^2/A]^{\frac{1}{2}}$. It follows that

$$f_x = \frac{1}{2}\left[\frac{\pi x^2}{A}\right]^{\frac{1}{2}} = \frac{1}{2}\left[\frac{8\pi m x^2 kT}{h^2}\right]^{\frac{1}{2}}$$

$$= \left[\frac{2\pi m kT}{h^2}\right]^{\frac{1}{2}} x.$$

Similar equations may be obtained for f_y and f_z, so that

$$f_t = \left[\frac{2\pi m kT}{h^2}\right]^{\frac{3}{2}} xyz \tag{5.4}$$

or

$$f_t = \left[\frac{2\pi m kT}{h^2}\right]^{\frac{3}{2}} V. \tag{5.5}$$

The physical properties of a gas do not depend on the *shape* of the container, so that (5.5) applies whatever the shape. Hence

$$\left(\frac{\partial \ln f_t}{\partial T}\right)_V = \frac{3}{2T}. \tag{5.6}$$

It follows from (4.40) that the translational contribution to the energy of an assembly of N molecules at temperature T is given by the equation

$$E_t^T = NkT^2\left(\frac{\partial \ln f_t}{\partial T}\right)_V = \frac{3}{2}NkT. \tag{5.7}$$

As we shall see in Section 5.9, expressions for $E^T - E^0$ (where E^0 is the energy of the assembly at absolute zero) are particularly useful. It is not really permissible[2] to calculate E_t^0 the translational energy of the assembly at absolute zero from (5.7), but we can calculate the lowest translational

[1] The divergence of the dashed curve in Figure 5.1 from the horizontal is of course grossly exaggerated. Its divergence is appreciable only when n_x approaches the value 10^8.

[2] See the footnote on page 80.

energy permissible to the molecule by putting $n_x = n_y = n_z = 1$ in equation (4.6). For a hydrogen molecule in a cubic container of sides one centimetre, this proves to be 5×10^{-37} J, and the corresponding value for a heavier molecule is obviously smaller still, whereas the *mean* translational energy $\frac{3}{2}kT$ is about 6×10^{-21} J for $T = 300$ K. It is clear therefore that at absolute zero the translational energy is *effectively* zero, (i.e. the molecule is effectively at rest) and we can write

$$E_t^0 = 0 \qquad (5.8)$$

so that

$$E_t^T - E_t^0 = \tfrac{3}{2}NkT. \qquad (5.9)$$

It follows from (4.41) that the translational contribution to the entropy of the assembly is given by the equation

$$S_t^T = Nk \ln f_t/N + NkT\left(\frac{\partial \ln f_t}{\partial T}\right)_V + Nk$$

$$= Nk \ln\left[\frac{(2\pi mkT)^{\frac{3}{2}}V}{Nh^3}\right] + \tfrac{5}{2}Nk. \qquad (5.10)$$

For some purposes it is more convenient to express the translational contribution to the entropy of a gas in terms of its molar mass M rather than m, and in terms of its pressure rather than its volume. The first is achieved by substituting M/L for m, and the second (anticipating the result to be obtained in Section 5.6) by substituting NkT/P for V, so giving the expression

$$S_t^T = Nk \ln\left[\frac{(2\pi M)^{\frac{3}{2}}(kT)^{\frac{5}{2}}}{L^{\frac{5}{2}}h^3 P}\right] + \tfrac{5}{2}Nk. \qquad (5.11)$$

The translational contribution to the *molar* entropy of a gas at 298.15 K and a pressure of one atmosphere may now be obtained from (5.11) by inserting the numerical values for L, h, k, T and P. The result[1] is

$$S_t(298.15, 1 \text{ atm}) = \tfrac{3}{2}R \ln M + 108.784 \text{ JK}^{-1} \text{ mol}^{-1}. \qquad (5.12)$$

5.2   The Rotational Partition Function

The rotational energy of a macroscopic rigid body of any shape may be resolved into three components referred to three orthogonal (i.e. mutually

---

[1] The passage from (5.11) to (5.12) should be verified by the reader. If k is expressed in JK^{-1} and h in Js, P must be expressed as 101.325 kNm^{-2}.

perpendicular) axes passing through its centre of mass. We can position the axes where we will, but the mechanics is simpler if we choose three particular axes, called the principal axes, about each of which the body has a moment called a principal moment of inertia. Any such body has therefore three principal moments of inertia, all of which may be different, or two or all three the same dependent on its shape.

In principle the rotation of a rigid molecule (the assumption that a molecule is rigid follows from the assumption that its rotational and vibrational activities are independent) is governed by the same considerations, but there are in some cases important restrictions. Consider a diatomic molecule A–B consisting of an atom A of mass m_A and an atom B of mass m_B separated by a distance r. One of the principal axes, axis 1 passes through the line of centres, axes 2 and 3 passing through the centre of mass P, but at right-angles to each other and to axis 1 as shown in Figure 5.2, axis 3 being supposed to be perpendicular to the plane of the page. The fact that we have assumed that the rotational activity of a molecule is separable from its electronic behaviour is permissible because the mass of the electrons in the molecule is so much smaller than that of the nuclei that the dynamic activity of the latter may be treated as though the electrons were not present (the Born–Oppenheimer principle). This means that we may treat the molecule as though the atoms A and B are point masses. The consequence of this is that to speak of the rotation of the molecule about axis 1 is *meaningless*, as we have no means of detecting whether it is rotating or not. Since we have no reason to suppose that rotation about axis 1 occurs at all we must account for the rotation of the molecule by considering only rotation about axes 2 and 3.

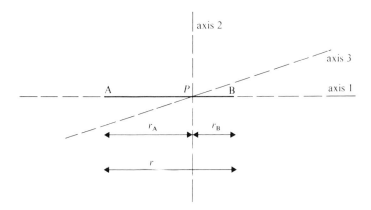

Figure 5.2   The principal axes of a diatomic molecule.

Precisely the same considerations show that rotation of a single atom (i.e. that of a single point mass) is meaningless. We conclude that rotation plays no part in the behaviour of a monatomic gas (or of atoms in an atomic crystal) and hence contributes nothing to the thermodynamic properties of such assemblies. We now return to the consideration of the rotation of a diatomic molecule about axes 2 and 3. The moments of inertia about these axes are clearly equal and are given by the equation

$$I_2 = I_3 = m_A r_A^2 + m_B r_B^2 \tag{5.13}$$

or identically, but more conveniently by the equation

$$I_2 = I_3 = \mu r^2 \tag{5.14}$$

where μ the so-called reduced mass is defined by the equation

$$1/\mu = 1/m_A + 1/m_B. \tag{5.15}$$

Thus the value of I_2 for the carbon monoxide molecule for which $m_C = 19.92 \times 10^{-27}$ kg, $m_0 = 26.56 \times 10^{-27}$ kg and $r = 1.13 \times 10^{-10}$ m is 14.5×10^{-47} kg m^2. The rotation of a linear polyatomic molecule is similar to that of a diatomic. We show in Figure 5.3 a linear triatomic molecule A–B–C, choosing again to place axis 1 through the line of centres of the three nuclei. Rotation about axes 2 and 3 is governed by the moments of inertia given by the equation

$$I_2 = I_3 = m_A r_A^2 + m_B r_B^2 + m_C r_C^2 \tag{5.16}$$

the distances r_A, r_B and r_C being given by the equations

$$m_A r_A + m_B r_B = m_C r_C \quad \text{and} \quad r_A + r_C = r.$$

Rotation about axis 1 is not possible.

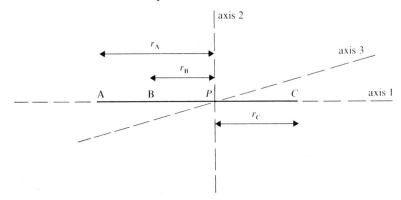

Figure 5.3   The principal axes of a linear triatomic molecule.

Figure 5.4 illustrates the rotation of the non-linear molecule A–B–A. We place axis 1 to coincide with the axis of symmetry (we will accept without proof that this makes axis 1 a principal axis,) and the other two at right angles to it. It is sufficient to note that all three moments are of the same order, and all must therefore be taken into account. We are now in a position to obtain equations for rotational energies and for the rotational partition function.

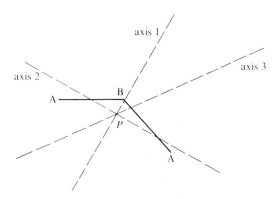

Figure 5.4   The principal axes of a non-linear molecule.

(i) *Hetero-nuclear diatomic and unsymmetrical linear polyatomic molecules*

Quantum theory shows that all diatomic and linear polyatomic molecules can assume values for the rotational energy given by the equation

$$\epsilon_r = \frac{h^2}{8\pi^2 I} J(J + 1) \tag{5.17}$$

where $I = I_2 = I_3$,[1] and where J is the rotational quantum number. In the case of a hetero-nuclear diatomic molecule A–B (the two atoms A and B being of unlike kind) or an unsymmetrical linear polyatomic molecule such as A–A–B or A–B–C, the quantum number J can assume any integral value from zero to infinity, each rotational level being $(2J + 1)$-fold degenerate.

The rotational partition function for such molecules is therefore given by the equation

$$f_r = \sum_{J=0}^{J=\infty} (2J + 1) \exp\left[-\frac{h^2}{8\pi^2 I k T} J(J + 1) \right]. \tag{5.18}$$

---

[1] When we speak of the moment of inertia of a linear molecule we mean I_2 or its identical twin I_3. In the case of a non-linear molecule we must give all three values I_1, I_2 and I_3.

We now define a quantity θ_r by the equation

$$\theta_r = \frac{h^2}{8\pi^2 Ik} \tag{5.19}$$

and so write

$$\epsilon_r = k\theta_r J(J + 1) \tag{5.20}$$

and

$$f_r = \sum_{J=0}^{J=\infty} (2J + 1) \exp\left[-J(J + 1)\theta_r/T\right]. \tag{5.21}$$

The quantity θ_r which has the dimensions of temperature and is called the characteristic rotational temperature, is better regarded more as a convenient parameter than as the measure of a quantity having a precise quantitative significance. It should be noted that in some publications θ_r is defined so that it has a value twice that of the quantity used here. Our next task is to show how the sum given in (5.21) may be evaluated.

It will be seen from Table 5.2 that the values of θ_r for diatomic molecules other than those containing hydrogen are very low indeed, thus θ_r for CO is 2.77 K, θ_r for NO is 2.42 K and so on, and it is obvious that the values of θ_r for linear polyatomic molecules are even lower. For such temperatures that θ_r/T is very much smaller than unity it is permissible to replace the sum on the right-hand side of (5.21) by the integral[1]

$$\int_{J=0}^{J=\infty} (2J + 1) \exp\left[-J(J + 1)\theta_r/T\right] dJ.$$

This integral is easily evaluated. If we replace $J(J + 1)$ by y, then $(2J + 1)\,dJ$ may be replaced by dy, so that

$$f_r = \int_0^\infty \exp\left(-y\theta_r/T\right) dy = -\frac{T}{\theta_r}\left[\exp\left(-y\theta_r/T\right)\right]_0^\infty$$

$$= T/\theta_r = \frac{8\pi^2 IkT}{h^2}. \tag{5.22}$$

This equation holds only if T is very much greater than θ_r, and cannot be used to obtain the correct value for f_r at 'low' temperatures. At temperatures equal to or below θ_r, f_r must be evaluated by direct summation, but this is not very tedious because all but three or four terms can be neglected. Thus for

---

[1] The justification for this replacement is similar to that given for the similar replacement in Section 5.1.

$$T = \theta_r,$$

$$f_r = 1 + 3 \exp(-2) + 5 \exp(-6) + 7 \exp(-12) + \cdots$$
$$= 1 + 0.4059 + 0.0124 + 0.0000_4 + \cdots$$
$$= 1.4183.$$

This procedure may of course be adopted at all temperatures, but for temperatures which are higher than θ_r but insufficiently high to permit the use of (5.22) it is usual to use an approximation formula due to Mulholland, i.e.

$$f_r = \frac{T}{\theta_r}\left[1 + \frac{1}{3}\left(\frac{\theta_r}{T}\right) + \frac{1}{15}\left(\frac{\theta_r}{T}\right)^2 + \frac{4}{315}\left(\frac{\theta_r}{T}\right)^3 + \cdots \right] \qquad (5.23)$$

The question 'how high does the temperature have to be so that equation (5.22) may be used without significant error?' is best answered by studying Tables 5.1 and 5.2. The first gives the values of f_r at various temperatures calculated according to the three methods given above, whilst Table 5.2 gives the values of θ_r for some diatomic molecules.

Table 5.1 Values of f_r calculated (i) by direct summation, (ii) from equation (5.23) and (iii) from equation (5.22).

T	f_r		
	(i)	(ii)	(iii)
20_r	2.370	2.367	2
100_r	10.343	10.340	10
200_r	not evaluated	20.336	20
500_r	not evaluated	50.33	50
1000_r	not evaluated	100.33	100

Table 5.2 Values of θ_r for some diatomic molecules.

	θ_r/K		θ_r/K
HD	64.0	H_2	85.4
HF	30.3	D_2	42.7
HCl	15.2	N_2	2.86
HBr	12.1	Cl_2	0.346
HI	9.0	Br_2	0.116
CO	2.77	I_2	0.054
NO	2.42	O_2	2.07

Realising that the moments of inertia of most linear polyatomic molecules are greater than those for most diatomic molecules, and the values for θ_r correspondingly less, and remembering that all expressions for the rotational contribution to the thermodynamic properties of gases contain terms in $\ln f_r$ rather than in f_r itself, it is evident that equation (5.22) may be used without sensible error for temperatures as low as 100 K for all unsymmetrical molecules except HD, HF, HCl, HBr and HI. (We shall consider the case of homonuclear diatomic molecules later.) The fact that the characteristic rotational temperature for most molecules is so low means that the percentage in excited rotational states is quite high even at low temperatures. The distribution of hetero–nuclear diatomic molecules over rotational energy levels at three very low temperatures, $\theta_r/2$, θ_r and $2\theta_r$ is given in Table 5.3, and that at three higher temperatures, $6.5\theta_r$, $19.5\theta_r$ and $65\theta_r$ shown graphically in Figure 5.5. The calculations on which the figure is based are those for $^1H^{35}Cl$, for which the characteristic rotational temperature is 15.37 K, so that $6.5\theta_r$ is 100 K, $19.5\theta_r$ is 300 K and $65\theta_r$ is 1000 K. The values of f_r required in these calculations were obtained by term-by-term summation.[1]

Table 5.3   Distribution numbers among rotational energy levels for hetero–nuclear diatomic molecules at very low temperatures.

T	n_0/N	n_1/N	n_2/N	n_3/N	n_4/N
$\frac{1}{2}\theta_r$	0.948	0.052	—	—	—
θ_r	0.705	0.286	0.009	—	—
$2\theta_r$	0.422	0.465	0.105	0.007_3	0.000_2

(ii) *Homonuclear diatomic and symmetrical linear polyatomic molecules*

A homonuclear diatomic molecule is of formula A–A, and a symmetrical linear polyatomic molecule of formula A–B–A or A–B–B–A, A and B representing atoms of different species. The rotational activity of such molecules depends on the relative orientations of the nuclear magnetic moments of the constituent atoms and is governed by quantum–mechanical selection rules. We shall investigate some of the consequences of these rules in Section 5.7, but both there and here we shall accept without proper justification the fact that such molecules can assume only *even* values of J, i.e. 0, 2, 4, ... or *odd* values 1, 3, 5, ... but not both. It follows that the rotational

---

[1] I am indebted to my colleague Dr. S. J. Moss for the calculations on which Figure 5.5 is based.

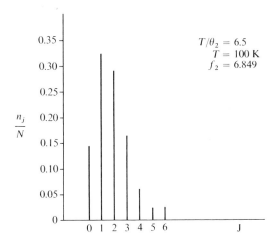

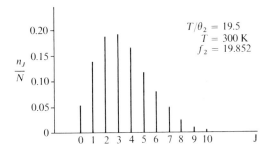

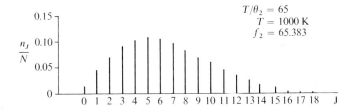

Figure 5.5 Distribution among rotational levels for $^1H^{35}Cl$; $\theta_r = 15.37$ K.

partition function for such molecules is given *not* by equation (5.21) but either by the equation

$$f_r' = \sum_{J=0,2,...} (2J + 1) \exp\left[-J(J + 1)\theta_r/T\right] \tag{5.24}$$

or by the equation

$$f_r'' = \sum_{J=1,3,...} (2J + 1) \exp\left[-J(J + 1)\theta_r/T\right]. \tag{5.25}$$

At temperatures very much greater than θ_r when many rotational states contribute to the partition function it is evident that

$$f_r' \simeq f_r'',$$

so that

$$f_r \simeq f_r' \simeq f_r'' \simeq \frac{1}{2} \sum_{J=0}^{J=\infty} (2J + 1) \exp\left[-J(J + 1)\theta_r/T\right] \tag{5.26}$$

which is of course the same as equation (5.22) except for the factor $\frac{1}{2}$. It follows that the high-temperature expression for f_r is

$$f_r = \frac{T}{2\theta_r} = \frac{8\pi^2 IkT}{2h^2}. \tag{5.27}$$

Quantum mechanics apart, it is not difficult to see why the rotational partition function for symmetrical linear molecules should, at high temperatures, be represented by equation (5.27) rather than by (5.22) because in statistical mechanics we are primarily concerned with the number of distinguishable complexions for the assembly, and molecular rotation contributes to the properties of the system simply because it contributes to that number. If a linear molecule is unsymmetrical each orientation A–A–B or B–A–A is distinguishable, but if symmetrical, two otherwise distinct orientations A–B–A' or A'–B–A are indistinguishable because atom A is indistinguishable from atom A', so that the total number of distinguishable complexions is one half that which it would otherwise be. A similar situation arises in the case of some non-linear polyatomic molecules. All distinct orientations of the molecule

$$\begin{array}{c} \text{O} \\ \diagup \quad \diagdown \\ \text{H} \qquad \text{D} \end{array}$$

are distinguishable, but only one half of the otherwise distinct orientations of the molecule

$$\begin{array}{c} \text{O} \\ \diagup \quad \diagdown \\ \text{H} \qquad \text{H} \end{array}$$

and one third of the otherwise distinct orientations of the molecule NH_3. It is convenient to ascribe to any molecule a *symmetry number* σ, formally defined as the number of otherwise distinct orientations which are indistinguishable due to the indistinguishability of like atoms. Thus heteronuclear diatomic and unsymmetrical linear polyatomic molecules are said to have a symmetry number of *one*, and homonuclear diatomic and symmetrical linear poly-atomics of *two*. The symmetry number for non-linear molecules must be obtained by inspection of the structural formula for the molecule. Thus NH_3 has a symmetry number of *three*, CH_4 a symmetry number of *twelve*, C_2H_4 of *four*, C_6H_6 of *twelve*.

The rotational partition function for *all* linear molecules, whether symmetrical or not, at high temperatures is therefore given by the equation

$$f_r = \frac{T}{\sigma\theta_r} = \frac{8\pi^2 I k T}{\sigma h^2} \tag{5.28}$$

σ having the value of *two* in one case and *one* in the other.

It is evident from the data in Table 5.2 that equation (5.28) may be used without sensible error to calculate the partition function for all homonuclear diatomic molecules except H_2 and D_2 at all temperatures with which we are likely to be concerned. The exceptional behaviour of H_2 and D_2 at very low temperatures is considered in Section 5.7.

We may now without further ado obtain expressions for the rotational contributions to the energy and entropy of an assembly of N identical diatomic or linear polyatomic molecules at temperatures high enough to permit the use of equation (5.28). From (5.28)

$$\ln f_r = \ln T/\sigma\theta_r \tag{5.29}$$

and

$$\frac{d \ln f_r}{dT} = \frac{1}{T}, \tag{5.30}$$

so that the rotational energy of the assembly at temperature T is given by the equation

$$E_r^T = NkT^2\frac{d \ln f_r}{dT} = NkT. \tag{5.31}$$

At absolute zero the rotational energy is zero so that

$$E_r^T - E_r^0 = NkT. \tag{5.32}$$

The rotational contribution to the entropy of the assembly is given by the equation

$$S_r^T = Nk \ln f_r + NkT\frac{d \ln f_r}{dT}$$

$$= Nk \ln \frac{T}{\sigma\theta_r} + Nk. \tag{5.33}$$

Corresponding expressions for the energy and entropy of the assembly at temperatures so low that the use of equation (5.28) is not permissible may be obtained by evaluating f_r by the Mulholland formula or by term-by-term summation.

(iii) Non-linear molecules

It has already been pointed out that all three moments of inertia must be taken into account when assessing the rotational contribution to the properties of an assembly of non-linear molecules. We shall not attempt to derive the formula for the partition function. It is given by the equation

i.e.

$$f_r = \frac{\pi^{\frac{1}{2}}}{\sigma} \left[\frac{8\pi^2 I_1 kT}{h^2} \right]^{\frac{1}{2}} \left[\frac{8\pi^2 I_2 kT}{h^2} \right]^{\frac{1}{2}} \left[\frac{8\pi^2 I_3 kT}{h^2} \right]^{\frac{1}{2}}$$

$$f_r = \frac{\pi^{\frac{1}{2}}}{\sigma} \left[\frac{T^3}{\theta_{r_1}\theta_{r_2}\theta_{r_3}} \right]^{\frac{1}{2}}, \tag{5.34}$$

θ_{r_1}, θ_{r_2} and θ_{r_3} being the three characteristic temperatures. It is of course obvious that for any polyatomic molecule all three moments are so large that the three characteristic temperatures are sufficiently low for equation (5.34) to be used without sensible error at all temperatures.[1] It will be noted too that we have included the symmetry number in the expression.

It follows from (5.34) that

$$\ln f_r = \tfrac{1}{2} \ln \frac{\pi T^3}{\theta_1\theta_2\theta_3} - \ln \sigma \tag{5.35}$$

and

$$\frac{d \ln f_r}{dT} = \frac{3}{2T} \tag{5.36}$$

so that the rotational contributions to the energy and entropy of an assembly of N identical non-linear molecules are given by the equations

$$E_r^T = E_r^T - E_r^0 = \tfrac{3}{2}NkT \tag{5.37}$$

and

$$S_r = \tfrac{1}{2}Nk \ln \frac{\pi T^3}{\theta_1\theta_2\theta_3} - Nk \ln \sigma + \tfrac{3}{2}Nk. \tag{5.38}$$

---

[1] This statement is not quite correct. The values of the moments for methane are so low that the partition function at very low temperatures should be evaluated by term-by-term summation.

(iv) *Internal rotation*

So far we have considered only the rotation of a rigid molecule about its centre of mass. In the case of some polyatomic molecules an additional rotational activity is possible, that of the rotation of one part of the molecule with respect to another. Thus in the case of toluene the methyl group may rotate with respect to the benzene ring

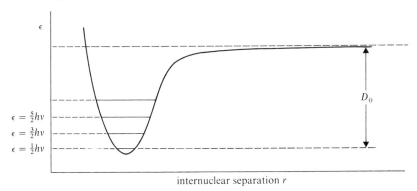

We shall postpone discussion of this particular mode of activity until Chapter 7, where we shall describe the part played by internal rotation in the chemistry of the xylenes.

5.3 The Vibrational Partition Function

When we speak of the vibrational activity of a non-localised molecule we mean the periodic variation in the distance between any two atoms therein. The phenomenon has therefore no meaning for a monatomic molecule.

(i) *Diatomic molecules*

The potential energy of a diatomic molecule A–B is a function of only one coordinate, the internuclear separation r, and is shown diagrammatically in Figure 5.6.

The vibrational activity is governed by the force constant λ, which for small amplitudes is proportional to r. In these circumstances the vibration

ϵ

$\epsilon = \frac{5}{2}h\nu$

$\epsilon = \frac{3}{2}h\nu$

$\epsilon = \frac{1}{2}h\nu$

D_0

internuclear separation r

Figure 5.6

is simple harmonic, and the energies are given by the equation

$$\epsilon_v = (v + \tfrac{1}{2})\frac{h}{2\pi}\left(\frac{\lambda}{\mu}\right)^{\frac{1}{2}} \tag{5.39}$$

where v, the vibrational quantum number, can assume any value $0, 1, 2, \ldots$, the quantity μ being the reduced mass given by the equation

$$\mu = \frac{m_A m_B}{m_A + m_B}. \tag{5.40}$$

The quantity $(1/2\pi)(\lambda/\mu)^{\frac{1}{2}}$ has the dimensions of frequency and is denoted by the symbol v, so that the energy levels may be described by the equation

$$\epsilon_v = (v + \tfrac{1}{2})hv, \quad v = 0, 1, 2, \ldots \tag{5.41}$$

the lowest possible vibrational energy being $\tfrac{1}{2}hv$. The quantity D_0 which is shown in Figure 5.6 represents the difference in potential energy of the molecule in its lowest vibrational state and that of the constituent atoms at infinite separation.

The vibrational energy levels are non-degenerate, so that the vibrational partition function for the diatomic molecule is given by the equation

$$f_v = \exp\left(-hv/2kT\right) + \exp\left(-3hv/2kT\right) + \exp\left(-5hv/2kT\right) + \cdots$$
$$= \frac{\exp\left(-hv/2kT\right)}{1 - \exp\left(-hv/kT\right)}. \tag{5.42}$$

We again introduce a parameter θ_v called the characteristic vibrational temperature, defined by the equation

$$\theta_v = \frac{hv}{k}, \tag{5.43}$$

so that

$$f_v = \frac{\exp\left(-\theta_v/2T\right)}{1 - \exp\left(-\theta_v/T\right)}, \tag{5.44}$$

$$\ln f_v = -\frac{\theta_v}{2T} - \ln\left[1 - \exp\left(-\theta_v/T\right)\right] \tag{5.45}$$

and

$$\frac{d \ln f_v}{dT} = \frac{\theta_v}{2T^2} + \frac{\theta_v/T^2}{\exp\left(\theta_v/T\right) - 1}. \tag{5.46}$$

It follows that the vibrational contribution to the energy of an assembly of N identical diatomic molecules at temperature T is given by the equation

$$E_v^T = NkT^2 \frac{d \ln f_v}{dT} = \tfrac{1}{2}Nk\theta_v + \frac{Nk\theta_v}{\exp(\theta_v/T) - 1}. \tag{5.47}$$

Obviously

$$E_v^0 = \tfrac{1}{2}Nk\theta_v \tag{5.48}$$

so that

$$E_v^T - E_v^0 = \frac{Nk\theta_v}{\exp(\theta_v/T) - 1}. \tag{5.49}$$

When considering chemical reactions between gases it is sometimes convenient to set the energy zero not at the base of the potential energy curve as we have done above, but to set it to coincide with the potential energy of the constituent atoms at infinite separation. The vibrational energies are then given by the expression

$$\epsilon_v = -D_0 + vh\nu, \quad v = 0, 1, 2, \ldots \tag{5.41'}$$

where D_0 is the difference in energy between the lowest vibrational level and the atoms separated at infinity. It then follows that

$$f_v' = \frac{\exp(D_0/kT)}{1 - \exp(-\theta_v/T)} \tag{5.44'}$$

$$\ln f_v' = \frac{D_0}{kT} - \ln[1 - \exp(-\theta_v/T)] \tag{5.45'}$$

$$\frac{d \ln f_v'}{dT} = -\frac{D_0}{kT^2} + \frac{\theta_v/T^2}{\exp(\theta^v/T) - 1}, \tag{5.46'}$$

$$E_v^{T'} = -ND_0 + \frac{Nk\theta_v}{\exp(\theta_v/T) - 1} \tag{5.47'}$$

$$E_v^{0'} = -ND_0 \tag{5.48'}$$

and

$$E_v^{T'} - E_v^{0'} = \frac{Nk\theta_v}{\exp(\theta_v/T) - 1}. \tag{5.49'}$$

This last quantity is the same as $E_v^T - E_v^0$ which is as it should be. The vibrational contribution to the entropy of the assembly is of course the same whether we choose to express the partition function by (5.44) or by (5.44'), and is given by the equations

$$S_v = Nk \ln f_v + NkT\frac{d \ln f_v}{dT} = Nk \ln f_v' + NkT\frac{d \ln f_v'}{dT}$$

$$= Nk\frac{\theta_v/T}{\exp(\theta_v/T) - 1} - Nk \ln[1 - \exp(-\theta_v/T)]. \qquad (5.50)$$

It will be realised that the equations for the vibrational energy and entropy of an assembly of gas molecules are the same (except for a factor of three) as those obtained in Section 3.2 for an assembly of *localised* atoms. There is however one important difference between the vibrational behaviour of an atom in a crystal and that in a non-localised molecule. This is that the force constant governing the vibrational activity of the molecule is so much greater than that operating in the case of the atom in the crystal (the energy required to break the covalent bond in the molecule being very much greater than the energy of sublimation per atom of the atomic solid) that the fundamental frequency and the parameter θ_v have values for most diatomic molecules about ten times greater than the corresponding quantities with which we were concerned in Chapter 3. This means of course that excited vibrational levels in

Table 5.4 Values of the characteristic vibrational temperature for diatomic molecules

	θ_v/K		θ_v/K		θ_v/K
H_2	6210	CO	3070	Cl_2	810
N_2	3340	NO	2690	Br_2	470
O_2	2230	HCl	4140	I_2	310

the gas molecule are occupied appreciably only at high temperatures. The values for θ_v for some diatomic molecules are given in Table 5.4. At temperatures very much less than θ_v,

$$\ln f_v \simeq -\frac{\theta_v}{2T} \quad \text{or} \quad \ln f_v' \simeq \frac{D_0}{kT}, \qquad (5.51)$$

$$\frac{d \ln f_v}{dT} \simeq \frac{\theta_v}{2T^2} \quad \text{or} \quad \frac{d \ln f_v'}{dT} \simeq -\frac{D_0}{kT^2}, \qquad (5.52)$$

$$E_v^T \simeq \tfrac{1}{2}Nk\theta_v \quad \text{or} \quad E_v^{T'} \simeq -ND_0, \qquad (5.53)$$

$$E_v^T - E_v^0 = E_v^{T'} - E_V^{0'} \simeq 0 \qquad (5.54)$$

and

$$S_v \simeq 0. \qquad (5.55)$$

(ii) *Polyatomic molecules*

In classical terms a molecule containing n atomic nuclei has $3n$ degrees of freedom. For a non-localised molecule three of these are associated with translational motion and two or three with rotational motion (depending on whether the molecule is linear or not) leaving $3n - 5$ or $3n - 6$ associated with molecular vibrations. Thus a non-linear tri-atomic molecule such as H_2O has three vibrational modes shown schematically as

(i)   O   H   H
(ii)   ←O   H   H
(iii)   O   H   H

whilst the linear tri-atomic molecule CO_2 has four

(i) ←O   C   O→

(iii) O→   C   O→

(ii) O   C   O

(iv) O   C   O $-$ $+$ $-$

the $+$ and $-$ signs referring to displacements perpendicular to the page.

Each vibrational mode is independent of the remainder and has its own fundamental frequency (although sometimes two or more have the same values as have (iii) and (iv) above in the case of CO_2) and its own characteristic temperature θ_v. The complete vibrational partition function for the molecule is therefore the product of $3n - 5$ or $3n - 6$ terms each of the form given in (5.44), and the total vibrational contribution to the energy and entropy of the assembly the sum of $3n - 5$ or $3n - 6$ expressions each of the form given in (5.47) and (5.50). At any particular temperature some may be excited and others not. Thus at 300 K the percentage of CO_2 molecules in excited vibrational modes corresponding to (i) above is small, large for (iii) and (iv) but negligible for (ii). It remains only to remark that in general the larger the number of atoms in the molecule the lower will be some values of θ_v, and so the greater will be the vibrational contributions to the properties of the gas.

5.4 The Electronic Partition Function

The electronic partition function for a molecule is given by the expression

$$f_e = g_{e_0} \exp(-\epsilon_{e_0}/kT) + g_{e_1} \exp(-\epsilon_{e_1}/kT(+ \cdots$$
$$= \exp(-\epsilon_{e_0}/kT)[g_{e_0} + g_{e_1} \exp(-\Delta \epsilon_{e_1}/kT) + \cdots] \qquad (5.56)$$

where $\Delta\epsilon_{e_1} = (\epsilon_{e_1} - \epsilon_{e_0})\ldots g_{e_0}, g_{e_1}, \ldots$ being the degeneracies of the various electronic energy levels. With very few exceptions the values of $\Delta\epsilon_{e_1}, \ldots$ are so much greater than the value of kT at temperatures with which the chemist is concerned that an atom or molecule is almost certain to be in its electronic ground state, so that

$$f_e = g_{e_0} \exp\left(-\epsilon_{e_0}/kT\right), \tag{5.57}$$

$$\ln f_e = \ln g_{e_0} - \epsilon_{e_0}/kT \tag{5.58}$$

and

$$\frac{d \ln f_e}{dT} = \epsilon_{e_0}/kT^2. \tag{5.59}$$

The electronic contributions to the energy and entropy of an assembly are therefore the same at all temperatures (including absolute zero) so that

$$E_e^T = NkT^2 \frac{d \ln f_e}{dT} = N\epsilon_{e_0}, \tag{5.60}$$

$$E_e^T - E_e^0 = 0 \tag{5.61}$$

and

$$S_e = Nk \ln f_e + NkT \frac{d \ln f_e}{dT} = Nk \ln g_{e_0}. \tag{5.62}$$

The ground state degeneracy is in fact unity for most molecules, notable exceptions being the alkali metals for which g_{e_0} is *two*, and oxygen molecules for which it is three. In most other cases the electronic contribution to the entropy of an assembly is zero.

Nitric oxide is one of the few molecules which is electronically excited at low temperatures. The lowest electronic energy level consists of a pair of states (so that g_{e_0} is two), and the first excited level (also consisting of a pair of states so that g_{e_1} is two) corresponds to a wave number only 121 cm^{-1} above the ground state, so that

$$\frac{\Delta\epsilon_{e_1}}{k} = \frac{hc\varpi}{k} = 174.3 \text{ K} \tag{5.63}$$

where c is the velocity of light. All excited energy levels other than the first are inaccessible except at *very* high temperatures, so that the electronic partition function is given by the equation

$$f_e = \exp\left(-\epsilon_{e_0}/kT\right)[g_{e_0} + g_{e_1} \exp\left(-174.3/T\right)]. \tag{5.64}$$

It follows that

$$E_e^T = N\epsilon_{e_0} + \frac{174.3Nk}{\exp(174.3/T) + 1}. \tag{5.65}$$

The corresponding expression for the electronic contribution to the entropy of nitric oxide is left as an exercise for the reader.

5.5   The Nuclear Partition Function

In principle we must accept, as stated in Sections 3.1 and 4.4 that each atom in a molecule can exist in any one of a number of nuclear energy levels $\epsilon_{n_0}, \epsilon_{n_1}, \ldots$ and can therefore be assigned a nuclear partition function

$$
\begin{aligned}
f_n &= g_{n_0} \exp(-\epsilon_{n_0}/kT) + g_{n_1} \exp(-\epsilon_{n_1}/kT) + \cdots \\
&= \exp(-\epsilon_{n_0}/kT)[g_{n_0} + g_{n_1} \exp(-\Delta\epsilon_{n_1}/kT) + \cdots].
\end{aligned} \tag{5.66}
$$

The values of $\Delta\epsilon_{n_1}, \ldots$ are however so high (at least of the order kT when T is 10^{10} K) that the chance of nuclear excitation at moderate temperatures is infinitesimal. It follows that we may write

$$f_n = g_{n_0} \exp(-\epsilon_{n_0}/kT), \tag{5.67}$$

$$\ln f_n = \ln g_{n_0} - \epsilon_{n_0}/kT \tag{5.68}$$

and

$$\frac{d \ln f_n}{dT} = \epsilon_{n_0}/kT^2, \tag{5.69}$$

so that the contributions to the energy and entropy of the assembly are

$$E_n^T = NkT^2 \sum \frac{d \ln f_n}{dT} = N \sum \epsilon_{n_0}, \tag{5.70}$$

and

$$S_n = Nk \sum \ln g_{n_0}, \tag{5.71}$$

there being of course one term in the last two equations for each atom in the molecule. Equation (5.70) holds for all temperatures (including absolute zero) below the temperature of nuclear excitation, so that

$$E_n^T - E_n^0 = 0. \tag{5.72}$$

The nuclear energy states of an atom are conserved during any physical process or chemical reaction, so that ΔS_n is zero for any such process. It is therefore customary to ignore nuclear states when obtaining values for all thermodynamic quantities, in other words to carry out all calculations as though the complete partition function were given by the equation

$$f = f_t \cdot f_r \cdot f_v \cdot f_e. \tag{5.73}$$

This notwithstanding it will be found in Section 5.7 when considering the behaviour of some homonuclear molecules at very low temperatures, that although we may ignore ground state nuclear spin degeneracies (which are a part of the term g_{n_0} appearing above) as far as calculations are concerned, the *influence* of such degeneracies is felt, because they determine the rotational levels accessible to such molecules.

5.6   The Pressure of the System

Since classical thermodynamics shows that $P = -(\partial A/\partial V)_T$ it follows from (4.34) that

$$P = NkT \left(\frac{\partial \ln f}{\partial V} \right)_T. \tag{5.74}$$

The volume of the system enters the expression for the partition function only through the translational partition function, so that it follows from (5.5) that

$$P = NkT/V. \tag{5.75}$$

The pressure of n moles of gas occupying volume V at temperature T is also given by the equation

$$P = nRT/V.$$

Since $n = N/L$, where L is the Avogadro constant, it follows that

$$k = R/L. \tag{5.76}$$

The Boltzmann constant k was identified as R/L in Chapter 1 by means of an argument which was persuasive, but which could hardly be described as rigorous. The argument put forward in the present section is conclusive.

5.7   Review of Results

We are now in a position to test the entire statistical–mechanical treatment of non-localised particles by comparing the predicted values for some of the thermodynamic properties of gases with those obtained by experiment. The most convenient properties to choose are C_v the molar heat capacity at constant volume, and the molar entropy.

(a) *The molar heat capacity at constant volume*

The molar heat capacity is the sum of contributions from each mode of activity of the molecule, i.e.

$$C_v = \left(\frac{\partial E}{\partial T}\right)_V = \left(\frac{\partial E_t}{\partial T}\right)_V + \frac{dE_r}{dT} + \frac{dE_v}{dT} + \frac{dE_e}{dT} \qquad (5.77)$$

the expressions for $E_t, E_r, \ldots$ being those obtained in earlier sections with N replaced by L the Avogadro constant.

Only the translational activity contributes to the heat capacity of a monatomic gas at temperatures at which the atoms remain electronically un-excited, so that it follows from equations (5.7) and (5.76) that

$$C_v = \left(\frac{\partial E_t}{\partial T}\right)_V = \tfrac{3}{2}Lk = \tfrac{3}{2}R \qquad (5.78)$$

For a diatomic molecule at such temperatures that the molecule is rotation-ally excited but neither vibrationally nor electronically excited, it follows from (5.7) and (5.31) that

$$C_v = \left(\frac{\partial E_t}{\partial T}\right)_V + \frac{dE_r}{dT} = \tfrac{3}{2}Lk + Lk = \tfrac{5}{2}R. \qquad (5.79)$$

The experimental values for C_v for monatomic and diatomic gases (except nitric oxide) around room temperature are precisely those predicted by the above equations. We would expect the value for C_v for a diatomic gas to fall to $3R/2$ at such temperatures that all molecules are in their lowest permissible rotational state, that is to say at temperatures well below the characteristic rotational temperature. As can be seen from Table 5.2 the only gases for which θ_r is sufficiently high for the change in C_v to be observed experimentally are H_2, D_2 and HD, and such observations have been made.

The translational and rotational contributions to the heat capacities of linear polyatomic molecules are the same as those given by equation (5.79), but it follows from (5.37) that the rotational contribution in the case of non-linear molecules is $3Lk/2$, i.e. $3R/2$. In addition however, most polyatomic molecules are appreciably excited vibrationally at low temperatures. We will illustrate this by calculating the molar heat capacity of carbon dioxide at 298.15 K. The four characteristic vibrational temperatures for carbon dioxide are given in Table 5.5 together with the corresponding values for

$$\left(\frac{\theta_v}{T}\right)^2 \frac{\exp(\theta_v/T)}{[\exp(\theta_v/T) - 1]^2}.$$

Table 5.5   The vibrational contributions to the heat capacity of CO_2 at 298.15 K

θ_v/K	1890	3360	954	954
θ_v/T at 298.15 K	6.34	11.27	3.20	3.20
$\dfrac{C_v}{Nk} = \left(\dfrac{\theta_v}{T}\right)^2 \dfrac{\exp(\theta_v/T)}{[\exp(\theta_v/T)-1]^2}$	0.0712	0.0016	0.4536	0.4536

It follows from (5.47) that each vibrational mode contributes

$$Lk\left(\frac{\theta_v}{T}\right)^2 \frac{\exp(-\theta_v/T)}{[\exp(\theta_v/T)-1]^2}$$

to the molar heat capacity, so that the total vibrational contribution is 0.98R, and the total heat capacity therefore $\frac{3}{2}R + R + 0.98R$ i.e. 3.48R in complete agreement with the experimental value.

(b) *The molar entropies of gases*

Predicted values for the molar entropies of some gases at 298.15 K and 1 atm pressure are shown as S_{stat} in Table 5.6, the values shown being the sum of the translational contribution obtained from equation (5.12), the rotational contribution obtained where appropriate from equations (5.33) or (5.38), the vibrational contributions obtained where appropriate from equation (5.50), and in the case of oxygen, the electronic contribution $Lk \ln g_{e_0}$ where g_{e_0} is *three*, (the electronic contribution for the remaining gases is zero). Ground state nuclear spin degeneracies are of course ignored. Since the statistical–mechanical equations used are those for assemblies of *independent* particles, the values S_{stat} are those the substances would be expected to exhibit were they *perfect gases*. Shown for comparison are the so-called 'calorimetric values' S_{cal}, that is to say the quantities $S^{298.15} - S^0$ calculated from experimental heat capacities, heats of evaporation and fusion and where appropriate, the heats corresponding to any phase changes in the solid state, the values S_{cal} containing a correction for deviations from perfect gas behaviour calculated as described in Section 5.11. Details of calculations leading to values of S_{cal} are given in all texts on classical thermodynamics, but to remind the reader of the method the appropriate terms for ethylene are given below, using the experimental data due to Egan and Kemp.[1] Ethylene boils at 169.40 K and freezes at 103.95 K and exhibits no

[1] Egan and Kemp, *J. Amer. Chem. Soc.* **59**, 1264 (1937).

phase change in the solid state, so that

$$S_{\text{cal}}^{298.15} = \sum_{0}^{298.15} qd \ln T$$

$$= \int_{169.40}^{298.15} C_{p(g)}d \ln T + \frac{L_e}{169.40} + \int_{103.95}^{169.40} C_{p(\text{liq})}d \ln T + \frac{L_f}{103.95}$$

$$+ \int_{15}^{103.95} C_{p(s)}d \ln T + \tfrac{1}{3}C_p^{15} + \Delta S_{\text{pgc}}. \qquad (5.80)$$

The origin of the first five terms on the right, the sum of which gives $(S^{298.15} - S^{15})$ is clear. The origin of the sixth term requires some explanation. The lowest temperature at which the heat capacity of crystalline ethylene was measured was 15 K. In order to obtain the value $(S^{15} - S^0)$ it was assumed that at 15 K ethylene follows the Debye limiting equation $C_p = \alpha T^3$ (where α is a proportionality constant) so that

$$\int_0^{15} C_{p(s)}d \ln T = \int_0^{15} \alpha T^2 \, dT = \left[\frac{\alpha T^3}{3}\right]_0^{15} = \tfrac{1}{3}C_p^{15}. \qquad (5.81)$$

The seventh term provides the necessary correction term for deviations from perfect gas behaviour at 298.15 K. The values for each term obtained by Egan and Kemp expressed in $JK^{-1} mol^{-1}$ are 21.42, 79.96, 33.14, 32.22, 51.17, 1.04 and 0.62, so that

$$S_{\text{cal}}^{298.15} = 219.6 \, JK^{-1} \, mol^{-1}.$$

Table 5.6 Standard molar entropies of perfect gases at 298.15 K and 1 atm.

	$S_{\text{stat}}/JK^{-1} mol^{-1}$	$S_{\text{cal}}/JK^{-1} mol^{-1}$
Ne	146.23	146.5
N_2	191.59	192.0
O_2	205.14	205.4
HCl	186.77	186.2
HBr	198.66	199.2
HI	206.69	207.1
Cl_2	223.05	223.1
CO_2	213.68	213.8
SO_2	247.99	247.9
NH_3	192.09	192.2
CH_3Cl	234.22	234.1
C_2H_4	219.53	219.6
C_6H_6	269.28	269.7

The agreement between the two sets of figures in Table 5.6 is remarkable, particularly when one considers the diversity of the methods used to obtain them (and the data required). The values for S_{stat} were obtained essentially by counting the number of complexions accessible to the system at 298.15 K and 1 atm resulting from the translational, rotational, vibrational and electronic activities of the molecule, and require only values of the molecular mass, the appropriate moments of inertia, the vibrational frequencies and, in the case of oxygen the electronic ground state degeneracy. Those for S_{cal} were obtained from thermal measurements carried out on macroscopic quantities of material at temperatures between 298.15 K and absolute zero, and require *no* knowledge of its intimate structure. The agreement establishes of course the method by which the statistical values were obtained, but it also throws light on the *state* of the substances at temperatures close to absolute zero.

It will of course be realised that classical thermodynamics provides no relationship by which the *absolute* value for the entropy of a system may be determined, nor indeed *recognises* such an expression, only providing relationships through which entropy *changes* may be determined. The quantity $219.6 \, JK^{-1} \, mol^{-1}$ quoted above is the value for such an entropy change, being the calorimetric measure of the entropy 'acquired' by one mole of ethylene as the result of its passage from the crystalline state at absolute zero to the perfect gas state at 298.15 K and 1 atm pressure. If we choose to use the term 'absolute entropy', then $219.6 \, JK^{-1}$ is the calorimetric measure of the difference between the unknown and unobtainable 'absolute' value of the entropy at 298.15 K and the unknown and unobtainable 'absolute' value for the entropy at absolute zero.

Experiment[1] shows however that the entropy change for isothermal processes (such as chemical reactions) involving only 'perfect' crystalline bodies appears to become vanishingly small at temperatures approaching absolute zero. (This finding was originally known as the Nernst hypothesis, but today is taken as one statement of the so-called third law of thermodynamics.) This means of course that the entropies of all such crystalline bodies approach the same value at absolute zero. It is customary therefore conventionally to assign the value zero to the entropy of all 'perfect' crystalline bodies at absolute zero, whereupon the value $219.6 \, JK^{-1} \, mol^{-1}$ becomes the calorimetric measure of the 'conventional' entropy of ethylene at 298.15 K. It is pertinent to remark that the meaning of the word 'perfect' in this context is by no means well-defined, but it taken to imply that the crystal is in complete internal equilibrium.

[1] The experimental foundation of the Nernst hypothesis and its practical importance is described in all texts on classical thermodynamics. See for example Everdell, *Introduction to Chemical Thermodynamics*, (E.U.P. 1965, pp. 202–209).

The value S_{stat} is no less conventional. Although we have actually made use of expressions for the various partition functions for the molecule, the value for S_{stat} really depends on the equation $S_{stat}^T = k \ln W^T$, but we have not calculated the *absolute* value of W at temperature T, (we have not for example taken into account nuclear spin degeneracies, nor even considered other possible configurational degeneracies within the nuclei), but *the factor by which the absolute value of W at temperature T exceeds the absolute value of W at absolute zero.* In other words, we have calculated the quantity

$$k \ln W^T/W^0$$

by taking into account *all temperature-dependent sources of disorder.* The value S_{stat}^T is therefore a statistical measure of the entropy at temperature T relative to that in that possibly hypothetical state in which all the molecules in the assembly are in their lowest translational, rotational, vibrational and electronic states and in which all configurational disorder (possibly present at higher temperatures) has been removed, so that the assembly is in a unique (non-degenerate) state as far as these temperature-dependent modes of activity are concerned.

The most important conclusions which may be drawn from the fact that the values of S_{stat}^T and S_{cal}^T for the gases listed in Table 5.6 are the same within the limits of experimental error are:

(a) that these substances *do* approach such a unique (non-degenerate) state,

and (b) that in the statistical–mechanical treatment we have not inadvertently 'forgotten' any other temperature-dependent mode of activity, for had we done so we should have found cases in which the experimental value S_{cal} *exceeds* that of the statistical quantity. No such case as this is known.

It is however conceivable that when the temperature of a substance falls towards absolute zero during the experimental determination of S_{cal}, some element of disorder is 'trapped' within the system for one reason or another, so that the lowest calorimetric measurements are made on a system in a metastable state unable to lose this residual entropy. In this event we should expect to find that the calculated value S_{stat} *exceeds* the experimental value S_{cal}. The results shown in Table 5.7 shows that such cases are known. The accepted explanation for the residual entropy for each of the first five substances listed is that it is due to configurational disorder. We will consider in detail only the case of carbon monoxide. A state of complete

Table 5.7   Standard molar entropies of perfect gases at 298.15 K and 1 atm
pressure revealing residual entropies at absolute zero

	S_{stat}/JK^{-1}	S_{cal}/JK^{-1}	$S_{stat} - S_{cal}$	
CO	197.95	193.3	4.65	$R \ln 2 = 5.77$
N_2O	219.99	215.1	4.89	$R \ln 2 = 5.77$
NO	211.00	207.9	3.10	$\frac{1}{2}R \ln 2 = 2.88$
H_2O	188.72	185.3	3.42	$R \ln (\frac{3}{2}) = 3.37$
D_2O	195.23	192.0	3.23	$R \ln (\frac{3}{2}) = 3.37$
H_2	130.66	124.0	6.66	$\frac{3}{4}R \ln 3 = 6.85$
D_2	144.85	141.8	3.05	$\frac{1}{3}R \ln 3 = 3.04$

configurational order in the carbon monoxide crystal would be one in
which all molecules exhibit the same orientation, i.e.

$$CO \quad CO \quad CO \quad CO \quad CO$$

$$CO \quad CO \quad CO \quad CO \quad CO.$$

The dimensions of the carbon and oxygen atoms are however sufficiently
alike that from the viewpoint of the energy a 'mixed crystal' in which some
molecules 'face right' and some 'face left' is only very little less stable than one
in which all molecules 'face right' or all 'face left.' When the crystal is first
formed (at its freezing point) it is therefore almost as likely to be in *any one*
disordered state such as

$$CO \quad OC \quad OC \quad CO \quad CO$$

$$CO \quad CO \quad OC \quad OC \quad CO$$

as in the most ordered arrangement shown above, and as the temperature
falls towards absolute zero the molecules are trapped in whatever position
they have taken up. A complete description of the crystal is therefore uncertain because each molecule may be in one of two orientations. The
maximum possible number of micro-states due to this possible configurational disorder accessible to a system containing L such molecules is therefore
2^L, and the corresponding entropy $k \ln 2^L$, i.e. 5.77 JK^{-1}. The fact that the
actual value for the residual entropy appears to be only 4.65 JK^{-1} is taken
to mean that the crystal is ordered to some degree. The explanation in the
case of nitrous oxide is the same. The molecule is linear, some molecules
'face right' $N{\equiv}N{=}O$, and some 'face left' $O{=}N{\equiv}N$. The closeness of the
value for the residual entropy of nitric oxide, 3.10 JK^{-1} to that of the value
$\frac{1}{2}R \ln 2$, i.e. 2.88 JK^{-1} suggests uncertainty in the orientation of one half
the total number of molecules (calculated as NO). It is therefore supposed
that at low temperatures nitric oxide forms the dimer $\begin{smallmatrix} N & O \\ O & N \end{smallmatrix}$ which can then

'face right' or 'face left.' Alternatively it may be, as suggested by Giauque and Johnston, that the dimer may exist in two isomeric forms having only slightly different energies. There is then an 'entropy of mixing' equal to $\frac{1}{2}k \ln 2$ per NO molecule, i.e. a molar entropy of mixing of $\frac{1}{2}R \ln 2$. The explanation for the residual entropy of ice is similar to that for CO and N_2O, but rather more complicated. A structure proposed by Pauling[1] shows a residual entropy of $R \ln \frac{3}{2}$ in close agreement with the experimental value of 3.42 JK^{-1}. That for deuterium oxide is explained by the same model.

The residual entropy of hydrogen ($6.66 \text{ JK}^{-1} \text{ mol}^{-1}$) and that of deuterium ($3.05 \text{ JK}^{-1} \text{ mol}^{-1}$) is due to the existence of ortho and para molecules. A complete description of this requires a much greater knowledge of quantum mechanics than is assumed in this book, and will not be attempted. The essence of the matter (presented without full quantum-mechanical justification) is as follows.

Any assembly of homonuclear molecules may be expected to be a mixture of two species distinguished by the relative orientations of their nuclear magnetic moments. If the orientations of the moments of the two atoms in the H_2 molecule are the same the molecule is called *ortho*-hydrogen, and if they are opposed the molecule is said to be *para*-hydrogen. In quantum mechanics only certain combinations of nuclear spin and rotational wave functions are allowed, molecules of *ortho*-hydrogen being permitted to assume only *odd* rotational quantum numbers $J = 1, 3, \ldots$, whilst molecules of *para*-hydrogen are permitted to assume only *even* values, $J = 0, 2, \ldots$. There are in fact *three* symmetric nuclear spin functions corresponding to *ortho*-hydrogen (in other words the nuclear spin degeneracy of the ground state is *three*), but only *one* anti-symmetric spin function associated with *para*-hydrogen (so that the nuclear spin ground state is non-degenerate). It follows that the combined rotational and nuclear spin partition function for H_2 (the molecules being assumed to be in their nuclear ground states) is given by the expression

$$f_{r,ns} = 3 \sum_{J=1,3,\ldots} (2J + 1)\exp\left[-J(J + 1)\theta_r/T\right]$$
$$+ 1 \sum_{J=0,2,\ldots} (2J + 1)\exp\left[-J(J + 1)\theta_r/T\right] \tag{5.82}$$

It follows from the Boltzmann distribution law that the *equilibrium* distribution of molecules between *ortho* and *para* at all temperatures is given by the equation

$$\left(\frac{N_0}{N_p}\right)_{H_2} = \frac{3 \sum_{J=1,3,\ldots} (2J + 1)\exp\left[-J(J + 1)\theta_r/T\right]}{1 \sum_{J=0,2,\ldots} (2J + 1)\exp\left[-J(J + 1)\theta_r/T\right]}. \tag{5.83}$$

[1] L. Pauling, *J. Amer. Chem. Soc.* **57**, 2680 (1935).

As explained in Section 5.2, at high temperatures the two summations $\sum_{J=1,3,\ldots}$ and $\sum_{J=0,2,\ldots}$ may be replaced by integrals, and both approach the value $T/2\theta_r$, so that at high temperatures

$$\left(\frac{N_o}{N_p}\right)_{H_2} \to 3 \tag{5.84}$$

which value is actually reached only slightly above room temperature, the distribution at 298.15 K being 2.99.

Equation (5.83) may be written in the form

$$\left(\frac{N_o}{N_p}\right)_{H_2} = \frac{3[3 \exp(-2\theta_r/T) + 7 \exp(-12\theta_r/T) + \cdots]}{[1 + 5 \exp(-6\theta_r/T) + \cdots]}$$

and it is seen that at temperatures approaching absolute zero all terms within the bracket in the numerator approach zero, as do all terms except unity in the denominator. It follows that at very low temperatures

$$\left(\frac{N_o}{N_p}\right)_{H_2} \to 0. \tag{5.85}$$

In other words the *equilibrium* state of a sample of hydrogen at very low temperatures would be one in which all molecules were *para*, their rotational behaviour being then governed by equation (5.24).

If during the experimental determination of S_{cal} the hydrogen were cooled down so slowly that the *equilibrium* mixture appropriate to each temperature were established, so that measurements of C_p were made at very low temperatures on pure *para*-hydrogen, the resultant value for S_{cal} would be very close indeed to the value for S_{stat}, i.e. 130.66 JK^{-1} mol^{-1}. It happens however that the conversion of *ortho*-hydrogen to *para* is so slow that the low-temperature measurements of C_p used in the calculation of S_{cal} are actually made not on the equilibrium mixture governed by equation (5.83) but on the metastable mixture, three-quarters *ortho* and one quarter *para*, appropriate only to high temperatures.[1] What is the rotational contribution to the entropy of such a mixture close to absolute zero? At such temperatures the rotation of the *para* molecules ceases because they can assume the rotational state $J = 0$, but the rotation of the *ortho* molecules persists because the lowest rotational quantum number they can assume is unity. The degeneracy of this level is *three*, so that each *ortho* molecule contributes an amount $k \ln 3$ to the entropy of the whole. The residual molar entropy of the metastable mixture containing one quarter *para* and three quarters *ortho* is therefore $\frac{3}{4}Lk \ln 3$, i.e. 6.85 JK^{-1}.

[1] Dennison, *Proc. Roy. Soc. A*. 115, 483 (1927).

The explanation of the residual entropy of deuterium is similar, differing only in the respect that the equilibrium distribution of nuclear spins at high temperatures here corresponds to two parts *ortho* and one part *para*, and in the fact that it is, in this case, the *para* molecule for which the lowest possible rotational level is that characterised by $J = 1$. The residual entropy is therefore $\frac{1}{3}Lk \ln 3$, i.e. 3.04 JK^{-1}.

The satisfactory explanation for the apparent residual entropies of hydrogen and deuterium raises the important question of why a similar 'discrepancy' between the statistical and calorimetric entropies is not observed in the case of other homonuclear molecules $^{16}O_2$, $^{14}N_2$, Cl_2, ... and some symmetrical polyatomic molecules.

The answer is simple, and has nothing to do with quantum–mechanical restrictions, being merely that whilst H_2 and D_2 molecules are able to rotate in the crystal at all temperatures (until they fall to rotational state $J = 0$) the rotation of all other molecules ceases (presumably for steric reasons) either on freezing or on passing into the crystalline form reached at lower temperatures. Evidence for this comes from values of entropies of fusion and entropy changes resultant from the transition of one crystalline form to another. If a molecule is able to rotate in the solid state as freely as in the liquid the molar entropy of fusion proves to be close to Lk (8.314 JK^{-1} mol^{-1}), that of H_2 being 8.36 JK^{-1} mol^{-1}, and that of D_2 being 10.45 JK^{-1} mol^{-1}. On the other hand, that[1] of Cl_2 is 37.2 JK^{-1} mol^{-1}, suggesting that all rotation ceases when the liquid freezes. The case of N_2 is slightly more complicated. The entropy of fusion is 11.42 JK^{-1} mol^{-1}, suggesting that the molecule is still rotating, to some extent at least, in the solid, *but* at 35.61 K N_2 undergoes a phase change to a different crystalline form,[2] losing a further quantity of entropy equal to 6.40 JK^{-1} mol^{-1}, suggesting that it is during this phase change that all rotational activity is finally 'frozen out.' (The case of N_2 is particularly interesting because quantum mechanics shows that if rotation down to lower temperatures were possible, the substance would show the same residual entropy as D_2, i.e. $\frac{1}{3}R \ln 3$.) The case for O_2 is similar. The entropy of fusion is small, suggesting that the molecule is still rotating in the first crystalline form reached on freezing, but phase changes occur at two lower temperatures during which all rotational activity ceases. (Even if this did not occur, the $^{16}O_2$ molecule would *not* in fact exhibit residual entropy. The nuclear spin degeneracy of the oxygen atom is unity, and it follows from quantum mechanics that there is only one symmetric nuclear spin function and *no* anti-symmetric spin functions. The single symmetric nuclear spin function can be coupled only with even-numbered J-rotational states, so that odd-numbered states are completely absent. At temperatures

---

[1] W. F. Giauque and T. M. Powell, *J. Amer. Chem. Soc.* **61,** 1970 (1939).
[2] W. F. Giauque and J. O. Clayton, *ibid* **55,** 4875 (1933).

close to absolute zero, all molecules are therefore in the lowest rotational state, $J = 0$, and no residual entropy is possible.)

5.8   Entropy Values and Third Law Calculations

The reader is reminded that one of the most important uses of entropy values lies in the calculation of entropy changes associated with chemical reactions. The use of the third law in such calculations requires that the entropy values for the reactants at the temperature of interest are those which correspond to the *same* value at absolute zero, so that ΔS^0 is zero as required by the Nernst hypothesis. This condition is satisfied by the entropy values quoted for those substances listed in Table 5.6 and for all substances for which S_{stat} and S_{cal} prove to be the same within the limits of experimental error, and in such cases it is obviously immaterial which value is used. The same is not true however for the substances listed in Table 5.7. It should be clear from the discussion given in the last section that a zero value for ΔS^0 for reactions in which these substances are concerned will be obtained only if the statistical values for the entropies are used.

5.9   Statistical–Mechanical Data Available
in the Literature

The most important thermodynamic properties of perfect gases listed in the literature are the molar entropies, the so-called *standard molar free energy function* $-(G_\circ^T - E^0)/T$, the term $G_\circ^T$ representing the value for the molar Gibbs function at temperature T and a standard pressure of one atmosphere, and the so-called *molar enthalpy function* $(H^T - E^0)/T$, (no specification of pressure being here necessary because the enthalpy of a perfect gas is independent of pressure). We have already shown how the entropy values are calculated, and we now show how values for the other two functions are obtained from the equations given in earlier sections.

The enthalpy function

The quantity H^T is the molar enthalpy of the gas at temperature T and is defined in classical thermodynamics by the equation

$$H^T = E^T + PV$$

and so, following equation (5.75) by the equation

$$H^T = E^T + LkT,$$

so that

$$\frac{H^T - E^0}{T} = \frac{1}{T}(E^T - E^0) + Lk. \tag{5.86}$$

The values for $E^T - E^0$ are the sum of the various contributions given in earlier sections. Thus:

(a) for a monatomic gas at temperatures at which excited electronic levels are unoccupied so that only translational contributions are significant,

$$E^T - E^0 = E_t^T - E_t^0$$

and so from equation (5.9)

$$\frac{H^T - E^0}{T} = \tfrac{3}{2}Lk + Lk = \tfrac{5}{2}R,$$

(b) for a diatomic gas at temperatures at which rotational levels are fully excited but vibrational and electronic excitation negligible,

$$E^T - E^0 = (E_t^T - E_t^0) + (E_r^T - E_r^0)$$

and so from equations (5.9) and (5.32)

$$\frac{H^T - E^0}{T} = \tfrac{3}{2}Lk + Lk + Lk = \tfrac{7}{2}R,$$

(c) for a diatomic gas which is both rotationally and vibrationally excited, it follows from (5.9), (5.32) and (5.49) that

$$\frac{H^T - E^0}{T} = \tfrac{3}{2}Lk + Lk + \frac{1}{T}\left(\frac{Lk\theta_v}{\exp(\theta_v/T) - 1}\right) + Lk,$$

and so on.

The standard molar free energy function

G, the Gibbs function or so-called Gibbs free energy of a system is defined in classical thermodynamics by the equations

$$G = E + PV - TS = A + PV.$$

It follows from (4.34) and (5.75) that

$$G^T = -NkT \ln \frac{f}{N}, \tag{5.87}$$

so that for a monatomic gas at temperatures at which electronic excitation is negligible, it follows from (5.5) and (5.58) that

$$G^T = -NkT \ln \frac{f_t f_e}{N}$$

$$= -NkT \ln \left[\frac{(2\pi m k T)^{\frac{3}{2}} V g_{e0}}{Nh^3} \right] + N\epsilon_{e0} \qquad (5.88)$$

and since $V/N = kT/P$ that

$$-\frac{G^T - E^0}{T} = Nk \ln \left[\frac{(2\pi m k T)^{\frac{3}{2}}}{h^3} \cdot \frac{kT}{P} \cdot g_{e0} \right]. \qquad (5.89)$$

The *standard* molar free energy function at temperature T and at a pressure of one atmosphere, may now be obtained from this equation by substituting L for N and, assuming that m is expressed in kg, k in JK^{-1} and h in Js, inserting the value 101.325 kNm^{-2} for P into the right-hand side.

The Gibbs function for an assembly of N diatomic molecules at temperatures at which they are rotationally and vibrationally excited is given by the expression

$$G^T = -NkT \ln \frac{f_t f_r f_v f_e}{N}.$$

It follows from the appropriate equations given in earlier sections that

$$G_0^T = -NkT \ln \left[\frac{(2\pi m k T)^{\frac{3}{2}} kT}{h^3 P^\dagger} \cdot \frac{T}{\sigma \theta r} \cdot \frac{g_{e0}}{[1 - \exp(-\theta_v/T)]} \right]$$

$$+ \tfrac{1}{2} Nk\theta_v + N\epsilon_{e0} \qquad (5.90)$$

where $P^\dagger$ is given the value 101.325 kNm^{-2}, and that

$$E^0 = \tfrac{1}{2} Nk\theta_v + N\epsilon_{e0} \qquad (5.91)$$

so that

$$-\frac{G_0^T - E^0}{T} = Lk \ln \left[\frac{(2\pi m k T)^{\frac{3}{2}} kT}{h^3 P^\dagger} \cdot \frac{T}{\sigma \theta_r} \cdot \frac{g_{e0}}{1 - \exp(-\theta_v/T)} \right]. \qquad (5.92)$$

The formula for polyatomic molecules differs from (5.92) only in so far that the term $[1 - \exp(-\theta_v/T)]$ must be replaced by the product of the appropriate number of similar terms, and if the molecule is non-linear, by the replacement of the term $T/\sigma\theta_r$ by the term $(1/\sigma)(\pi T^3/\theta_1\theta_2\theta_3)^{\frac{1}{2}}$ as required by equation (5.35). It will be noticed that the value for the free energy function is relatively insensitive to changes in temperature. Its value can therefore be tabulated at fairly large temperature intervals and its value at intermediate temperatures obtained by graphical interpolation without undue error. This is the reason why $-(G_o^T - E^0)/T$ is calculated rather than $-(G_o^T - E^0)$.

Values for the molar enthalpy function, and the standard free energy function of some substances calculated as perfect gases at various temperatures, together with the values for the molar heats of formation of the compounds at 298.15 K are given in Table 5.8, the values having been extracted from tables given in the National Bureau of Standards Circular No. C461 (1947), but recalculated to conform to S.I. units. Values for similar but not identical functions are also given in J.A.N.A.F. *Tables of Thermochemical Data*, published by the Dow Chemical Company, Midland, Michigan. In this publication values are given (over a very large temperature range) for

Table 5.8

Temperature	298.15 K	600 K	1000 K	1500 K
$\dfrac{H^T - E^0}{T}$ $\Big/$ $JK^{-1} mol^{-1}$				
O_2	29.04	29.84	31.37	32.85
H	20.79	20.79	20.79	20.79
H_2	28.40	28.79	29.16	29.83
H_2O	33.20	33.98	35.90	38.71
CO	29.09	29.36	30.36	31.68
CO_2	31.41	37.12	42.77	47.43
C_2H_2	33.56	42.73	50.58	57.30
C_6H_6	47.74	85.69	126.19	160.00
$-\dfrac{G_o^T - E^0}{T}$ $\Big/$ $JK^{-1} mol^{-1}$				
O_2	175.98	191.10	212.12	225.13
H	93.83	108.36	118.98	127.41
H_2	102.19	122.19	136.98	148.91
H_2O	155.53	178.94	196.72	211.80
CO	168.82	189.21	204.43	217.00
CO_2	182.23	206.02	226.39	244.69
C_2H_2	167.26	193.77	217.59	239.46
C_6H_6	221.46	266.52	320.37	378.44
$\Delta H_f^{298.15}/kJ\ mol^{-1}$				
O_2	0			
H	+217.94			
H_2	0			
H_2O	-241.83			
CO	-110.52			
CO_2	-393.51			
C_2H_2	+226.75			
C_6H_6	+82.93			

the functions $(H^T - H^{298.15})/T$ and $-(G_0^T - H^{298.15})/T$. A most valuable collection of recent values for the thermodynamic properties of organic compounds is provided by Stull, Westrum and Sinke, *The Chemical Thermodynamics of Organic Compounds* (Wiley, 1969). The use of data such as those given in Table 5.8 and in *J.A.N.A.F. Tables* in calculating the standard free energy change and the corresponding equilibrium constants for gaseous reactions is described in Chapter 7.

5.10   Expressions for the Chemical Potential of a Perfect Gas

The molecular chemical potential of a pure substance is defined by the equation

$$\mu = G/N.$$

It follows from (5.88) that the chemical potential of a monatomic gas at temperatures at which electronic excitation is negligible is given by the expression

$$\mu = -kT \ln \left[\frac{(2\pi mkT)^{\frac{3}{2}} V g_{e_0}}{h^3 N} \right],$$

the chosen energy zero being that of the ground electronic state. Since

$$V/N = kT/P,$$

$$\mu = -kT \ln \left[\frac{(2\pi mkT)^{\frac{3}{2}} kT g_{e_0}}{h^3 P^{\dagger}} \right] + kT \ln \frac{P}{P^{\dagger}}, \tag{5.93}$$

the term $P^{\dagger}$ figuring twice on the right-hand side representing a chosen 'standard pressure' expressed in the same units as P, and introduced so that the argument of each logarithmic term remains dimensionless as it must. Equation (5.93) may be written in the form

$$\mu = \mu^{\dagger} + kT \ln \frac{P}{P^{\dagger}} \tag{5.94}$$

where

$$\mu^{\dagger} = -kT \ln \left[\frac{(2\pi mkT)^{\frac{3}{2}} kT g_{e_0}}{h^3 P^{\dagger}} \right]. \tag{5.95}$$

$\mu^{\dagger}$ is known as the 'standard' chemical potential of the gas, the word 'standard' relating to the chosen standard pressure $P^{\dagger}$. Thus, insertion of the value $P^{\dagger} = 1 Nm^{-2}$ into (5.95) gives the numerical value of $\mu^{\dagger}$ at a pressure of

$1 \mathrm{Nm}^{-2}$, whilst insertion of the value $P^\dagger = 101.325\ \mathrm{kNm}^{-2}$ gives the value of $\mu^\dagger$ at a pressure of one atmosphere.

It is perhaps appropriate to remind the reader that the chemical potential (partial *molar* free energy) occurring in classical thermodynamics and in this book denoted by the symbol μ^* is related to the molecular chemical potential by the equation $L\mu = \mu^*$. It follows that for a monatomic gas

$$\mu^* = -RT \ln \left[\frac{(2\pi mkT)^{\frac{3}{2}} kT g_{e_0}}{h^3 P^\dagger} \right] + RT \ln \frac{P}{P^\dagger}. \tag{5.96}$$

This equation is of course the statistical–mechanical foundation for the classical thermodynamic relation

$$\mu^* = \mu^{\dagger *} + RT \ln \frac{P}{P^\dagger}. \tag{5.97}$$

It follows from (5.90) that the chemical potential for a diatomic gas at temperatures at which it is both rotationally and vibrationally excited is given by the expression

$$\mu = -kT \ln \left[\frac{(2\pi mkT)^{\frac{3}{2}} kT}{h^3 P^\dagger} \cdot \frac{T}{\sigma\theta_r} \cdot \frac{g_{e_0}}{[1 - \exp(-\theta_v/T)]} \right]$$

$$+ kT \ln \frac{P}{P^\dagger}, \tag{5.98}$$

the chosen energy zero being now that of the ground electronic and lowest molecular vibrational state. As mentioned in Section 5.3 it is sometimes necessary to set the energy zero to coincide with the potential energy of the constituent atoms at infinite separation. The appropriate expression for μ is now obtained precisely as above except that the vibrational partition function is now described by (5.44') instead of (5.44). This leads to the expression

$$\mu = -kT \ln \left[\frac{(2\pi mkT)^{\frac{3}{2}} kT}{h^3 P^\dagger} \cdot \frac{T}{\sigma\theta_r} \cdot \frac{g_{e_0}}{[1 - \exp(-\theta_v/T)]} \right]$$

$$- D_0 + kT \ln \frac{P}{P^\dagger} \tag{5.99}$$

where D_0 is the difference in energy between the lowest vibrational level and the atoms separated at infinity. Expressions for the chemical potential of polyatomic molecules differ from (5.98) or (5.99) only in the rotational and vibrational terms, as has been explained in the previous section (see the paragraph following equation 5.92).

5.11   The Relation between the Entropy of a Perfect Gas and that of a Real (Imperfect) Gas

In Section 5.7 we compared the statistical–mechanical values for the molar entropies of perfect gases with the corresponding calorimetric values 'corrected for deviations from perfect gas behaviour'. We now show how such correction terms may be obtained. Alternatively, we may regard the contents of this section as the recipe for converting statistical–mechanical values for a perfect gas into those for a real (imperfect) gas.

All gases approach perfect gas behaviour when sufficiently rarified, and so if the values for the molar entropy of a real gas and its perfect counterpart are plotted against the pressure at constant temperature, the two curves would supposedly join at some point corresponding to some very low pressure P^*. Such a plot is shown diagrammatically in Figure 5.7, C being the imagined point of interception.

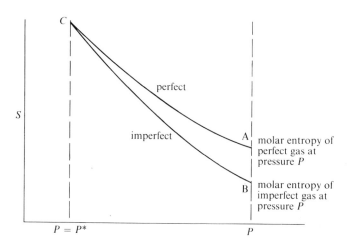

Figure 5.7

The quantity we are interested in is the difference between the molar entropy of a perfect gas at temperature T and pressure P, and that of the corresponding real gas at the same temperature and pressure. If we denote this quantity by δS, it is clear from Figure 5.7 that

$$\delta S = S_A - S_B = (S_A - S_C) - (S_B - S_C)$$

$$= \left[\int_{P^*}^{P}\left(\frac{\partial S}{\partial P}\right)_T dP\right]_{\text{perfect}} - \left[\int_{P^*}^{P}\left(\frac{\partial S}{\partial P}\right)_T dP\right]_{\text{imperfect}} \tag{5.100}$$

One of Maxwell's thermodynamic relations[1] between properties of systems of fixed composition is

$$\left(\frac{\partial S}{\partial P}\right)_T = -\left(\frac{\partial V}{\partial T}\right)_P,$$

so that

$$\delta S = \left[\int_{P*}^{P}\left(\frac{\partial V}{\partial T}\right)_P dP\right]_{\text{imperfect}} - \left[\int_{P*}^{P}\left(\frac{\partial V}{\partial T}\right)_P dP\right]_{\text{perfect}} \qquad (5.101)$$

For a perfect gas, $PV = RT$, so that $(\partial V/\partial T)_P = R/P$. It follows that

$$\delta S = \int_{P*}^{P}\left[\left(\frac{\partial V}{\partial T}\right)_{P\,\text{imperfect}} - \frac{R}{P}\right] dP. \qquad (5.102)$$

δS may be evaluated precisely from this equation if experimental P–V–T data are available so that $(\partial V/\partial T)_P$ can be evaluated for various pressures between P and P^*, and so expressed as a function of P.

If the necessary P–V–T data are not available, but the critical temperature and pressure of the gas are known, a value for δS may be estimated using the Berthelot equation of state

$$V = \frac{RT}{P} + \frac{9RT_c}{128P_c}\left[1 - 6\frac{T_c^2}{T^2}\right], \qquad (5.103)$$

from which it follows that

$$\left(\frac{\partial V}{\partial T}\right)_P = \frac{R}{P} + \frac{27RT_c^3}{32P_cT^3}.$$

Substituting into (5.102) and putting $P^* = 0$ for convenience, we see that

$$\delta S = \int_0^P \frac{27RT_c^3}{32P_cT^3} dP = \frac{27RP_r}{32T_r^3}, \qquad (5.104)$$

where $P_r(= P/P_c)$ and $T_r(= T/T_c)$ are the reduced pressure and temperature. The Berthelot equation is inexact, and its use in the present context has been criticised.[2] The position is that although equation (5.104) gives results for 'permanent' gases which are correct to within a few per cent, results for saturated vapours at temperatures at which P_{sat} is high may be too low by factors as great as five.

Values for δS are small for most gases and vapours at low pressures, those for $P = 1$ atm and $T = 298.15$ K for O_2, CO_2, HCl and NH_3 being 0.02,

[1] See Appendix I.

[2] Halford, *J. Chem. Phys.* **17**, 111, 405 (1945).

0.10, 0.11 and 0.16 JK^{-1} mol^{-1} respectively, and that for water vapour at 298.15 K and its saturated vapour pressure (0.024 atm) being 0.01 JK^{-1} mol^{-1}.

EXERCISES

5.1　One mole of oxygen O_2 is contained in a vessel volume V at 298.15 K and 1 atm pressure.

(a) using the formulae

$$f_t = \frac{(2\pi mkT)^{\frac{3}{2}}V}{h^3} = \frac{(2\pi m)^{\frac{3}{2}}}{h^3} \cdot \frac{L(kT)^{\frac{5}{2}}}{P^{\dagger}}$$

where $P^{\dagger} = 1$ atm $= 101.325$ kNm^{-2}, and $m = 32 \times 1.66 \times 10^{-27}$ kg, calculate $\ln f_t$.

(b) The internuclear distance in the O_2 molecule is 1.207×10^{-10} m. Calculate $\ln f_r$.

(c) The electronic ground state is three-fold degenerate, the electronic excitation at 298.15 K is negligible as is vibrational excitation. Calculate $\ln f_e$.

(d) Hence evaluate the standard molar entropy of O_2 at 298.15 K.

5.2　In most exercises concerning gases it is usually necessary to calculate

$$\frac{f}{N}, \quad \text{i.e.} \quad \frac{f_t}{N} \cdot f_r \cdot f_v \cdot f_e.$$

Show that for one mole of gas at one atmosphere pressure,

$$\frac{f_t}{L} = 0.02559 M^{\frac{3}{2}} T^{\frac{5}{2}}$$

where M is the 'conventional' dimensionless molecular weight.

5.3　(a) Use the values of θ_v/T for CO_2 at 298.15 K given in Table 5.5 to calculate the vibrational contribution to the molar entropy of CO_2 at that temperature. The formula required is

$$S_v = R \sum_{1,2,3,4} \left\{ \frac{\theta_v/T}{\exp(\theta_v/T) - 1} - \ln\left[1 - \exp(-\theta_v/T)\right] \right\}$$

the summation being extended over all four vibrational modes. The formula may be evaluated using mathematical tables, or by making use of tables of Einstein functions such as those given by Sherman and Ewell, *J. Phys. Chem.* **46**, 641, 1942.

(b) Calculate the value of S_t for one mole of gas at one atmosphere pressure and 298.15 K, (if you make use of the formula given in Exercise 5.2 remember to use the value $M = 44$), and the value of S_r (the CO_2 molecule is linear, so that $\sigma = 2$, and the value for θ_r is 0.564 K.)

(c) Hence evaluate the standard molar entropy of CO_2 at 298.15 K and check your answer by comparing it with the value given in Table 5.6.

5.4　The molecular weights of N_2 and CO are the same, the difference between their moments of inertia is negligible and neither molecule is vibrationally or electronically excited at 298.15 K. The standard molar entropy of the former is 191.5 JK^{-1} and that of the latter 197.9 JK^{-1}. Account for this difference.

5.5 The residual molar entropy at absolute zero of NO appears to be $\frac{1}{2}Lk \ln 2$, and that of CH_3D appears to be about $2Lk \ln 2$. What information do these figures give regarding the 'state' of NO and CH_3D at absolute zero?

5.6 The critical temperature for NH_3 is 405.6 K and the critical pressure 112.5 atm. Use these values and the Berthelot equation of state to justify the statement that the standard molar entropy of NH_3 at 298.15 K is 0.16 J K^{-1} less than the statistical value for the perfect gas.

CHAPTER 6

Perfect Gas Mixtures

In this chapter we obtain statistical–mechanical expressions for the thermo-dynamic properties of perfect gas mixtures and for the quantities of mixing, and most important of all, for the chemical potentials of components of such mixtures. We turn first however to the statistical–mechanical interpretation of the terms *heat* and *work*, because this provides an insight into the changes in molecular parameters which result from some of the processes with which we shall be concerned both in the present and in later chapters.

6.1 The Statistical–Mechanical Interpretation of Heat and Work

Since the energy of any assembly of independent particles of the same kind may be represented by the equation

$$E = \sum_i n_i \epsilon_i \tag{6.1}$$

it is clear that changes in E may result from shifts in the energy levels ϵ_i, *or* from shifts in the population numbers n_i, *or* from shifts in both, so that the change in energy resulting from any infinitesimal process is given by the equation

$$dE = \sum_i n_i \, d\epsilon_i + \sum_i \epsilon_i \, dn_i. \tag{6.2}$$

It was shown in Sections 3.2 and 3.4 that an increase in the specific volume V/N of a localised independent particle *decreases* its characteristic frequency ν, and so decreases the values of the permissible energy levels ϵ_i, and in Section 4.2 it was shown that the translational energy levels accessible to a non-localised particle are determined by the dimensions of the vessel to which it is confined, increase in the values of x, y and z (see equation 4.6) leading to a decrease in the values of ϵ_i. It follows in both cases, that if work is done on a system so decreasing its volume, an increase in the values of ϵ_i results. It appears therefore, that the term $\sum_i n_i\, d\epsilon_i$ figuring in (6.2) represents the change in energy of an assembly resulting from an infinitesimal change in volume, and so represents the work performed on the system by the external agency bringing about the volume change.

The classical statement of the first law of thermodynamics is that for a closed system

$$dE = q + w,$$

where q is the heat absorbed by the system, and w the work performed on it. The identification of w with $\sum_i n_i\, d\epsilon_i$ suggests the identification of q with $\sum_i \epsilon_i\, dn_i$. We suppose therefore that the absorption of heat by a system maintained at constant volume leads to changes in the population numbers but to no changes in the permissible energy levels. The absorption of heat by a system maintained at constant pressure will (supposing the process to be accompanied by an expansion of the system), result in shifts in both the energy levels and in the population numbers.

6.2   Mixtures of Independent Non-Localised Particles

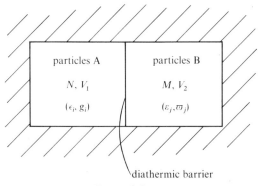

diathermic barrier

Figure 6.1

Consider first two sub-systems, one consisting of N independent non-localised particles A occupying volume V_1, and the second of M independent non-localised particles of another kind B occupying volume V_2. Suppose that they are separated, as shown in Figure 6.1, by a diathermic barrier through which they can exchange energy only as heat, but that the system as a whole is isolated. We require expressions for the thermodynamic properties for the system as a whole supposing that thermal equilibrium is established. We use the symbols ϵ_i, g_i and n_i to denote the energy levels accessible to particles A, their degeneracies and population numbers, and the symbols ε_j, ϖ_j and m_j to denote the corresponding quantities for particles B. We observe, from the findings of the last section, that passage of heat from one sub-system to another results in no changes in ϵ_i, g_i, ε_j and ϖ_j but only to changes in the population numbers n_i and m_j.

The population numbers n_i and m_j are governed by the two equations

$$N = \sum_i n_i \tag{6.3}$$

and

$$M = \sum_j m_j, \tag{6.4}$$

but the energy of the system by the single equation

$$E = \sum_i n_i \epsilon_i + \sum_j m_j \varepsilon_j. \tag{6.5}$$

The number of complexions accessible to the system as a whole must be the product of the numbers accessible to each sub-system (since any one complexion of one may be combined with any one of the other), so that, utilising equation (4.12), we see that for any set of population numbers n_i and m_j,

$$W = \prod_i \frac{g_i^{n_i}}{n_i!} \cdot \prod_j \frac{\varpi_j^{m_j}}{m_j!}. \tag{6.6}$$

The most probable sets of population numbers are therefore those which maximise W as given by (6.6), and so those which satisfy the equation

$$\delta \ln W = \sum_i \left(\frac{\partial \ln W}{\partial n_i}\right) \delta n_i + \sum_j \left(\frac{\partial \ln W}{\partial m_j}\right) \delta m_j = 0 \tag{6.7}$$

subject to the conditions

$$\delta N = \sum_i \delta n_i = 0, \tag{6.8}$$

$$\delta M = \sum_j \delta m_j = 0 \tag{6.9}$$

and

$$\delta E = \sum_i \epsilon_i \delta n_i + \sum_j \varepsilon_j \delta m_j = 0. \tag{6.10}$$

We now proceed by multiplying (6.8) by a multiplier α, (6.9) by a different multiplier α' (since condition 6.8 is independent of condition 6.9), but multiplying (6.10) by a single multiplier $-\beta$ (since the energy of one sub-system is *not* independent of that of the other), and on adding to (6.7) obtain the combined condition

$$\sum_i \left[\frac{\partial \ln W}{\partial n_i} + \alpha - \beta \epsilon_i \right] \delta n_i + \sum_j \left[\frac{\partial \ln W}{\partial m_j} + \alpha' - \beta \varepsilon_j \right] \delta m_j = 0. \tag{6.11}$$

We shall omit all further steps in the argument, because the procedure is similar to that adopted in Sections 2.1 and 4.2, and proceed to the final result, which is that

$$\ln W_{\text{max}} = N \ln \frac{f_A}{N} + N + M \ln \frac{f_B}{M} + M + \beta E \tag{6.12}$$

where the partition functions f_A and f_B are given by the equations

$$f_A = \sum_i g_i \exp(-\beta \epsilon_i) \tag{6.13}$$

and

$$f_B = \sum_j \varpi_j \exp(-\beta \varepsilon_j), \tag{6.14}$$

so that the entropy of the system is given by the equation

$$S = k \ln W_{\text{max}} = kN \ln \frac{f_A}{N} + kM \ln \frac{f_B}{M} + (N + M)k + \beta kE. \tag{6.15}$$

We may now show, using precisely the same method as that used in Section 2.5 that

$$\beta = \frac{1}{kT} \tag{6.16}$$

where T is the equilibrium temperature of the system as a whole.

Here we digress for a moment from the main purpose of this section to draw attention to the fact that a most important result has emerged almost imperceptibly from the procedure adopted. We refer to the fact that the variation in energy of the two-part system is described by a *single* equation (6.10) and so requires a single-valued multiplier $-\beta$ when proceeding to the combined condition given in (6.11). We see that two sub-systems in thermal equilibrium are described by the same value of β, and, since we have now identified β as $1/kT$, that *two sub-systems in thermal equilibrium are characterised by the same temperature*. In fact we have provided a statistical–mechanical foundation for the zeroth law of thermodynamics. We have of course considered only the case in which the two sub-systems are assemblies of independent non-localised particles, but in Chapter 8, when constructing the equations of the canonical ensemble, we shall, by using an identical argument obtain the same result for systems of all kinds.

For our present purpose we now make use of the fact that the partition function for all non-localised particles is proportional to the volume accessible to them, and so write

$$f_A = f'_A V_1 \quad \text{and} \quad f_B = f'_B V_2 \qquad (6.17)$$

where

$$f'_A = \frac{(2\pi m_A k T)^{\frac{3}{2}}}{h^3} \cdot (f_r f_v f_e)_A \qquad (6.18)$$

and

$$f'_B = \frac{(2\pi m_B k T)^{\frac{3}{2}}}{h^3} \cdot (f_r f_v f_e)_B \qquad (6.19)$$

and making use of (6.16) obtain the following equations:

$$S = kN \ln \frac{f'_A V_1}{N} + kM \ln \frac{f'_B V_2}{M} + (N + M)k + \frac{E}{T}, \qquad (6.20)$$

$$E = kNT^2 \left(\frac{\partial \ln f'_A V_1}{\partial T} \right)_{V_1} + kMT^2 \left(\frac{\partial \ln f'_B V_2}{\partial T} \right)_{V_2} \qquad (6.21)$$

and

$$A = -kNT \ln \frac{f'_A V_1}{N} - kMT \ln \frac{f'_B V_2}{M} - (N + M)kT. \qquad (6.22)$$

We now come to the final operation in the present section. We suppose that the diathermic barrier separating the two sub-systems is removed so that the two types of particles can mix.

From the statistical–mechanical viewpoint the only essential feature resulting from the removal of the barrier separating the two species is the fact that the total volume $V = V_1 + V_2$ is now accessible to all particles so that (through changes in the translational energy levels) the total energies ϵ_i and ε_j change, as do the corresponding degeneracies and population numbers. The final state of the system is therefore described by a set of equations which are precisely the same as those given above except that the terms $\epsilon_i, g_i, n_i, \varepsilon_j, \varpi_j, m_j$ and the partition functions f_A and f_B are replaced by the terms $\epsilon_i^*, g_i^*, n_i^*, \varepsilon_j^*, \varpi_j^*, m_j^*$ and the partition functions f_A^* and f_B^*. Furthermore, since the temperature is unchanged by the mixing process (the reader should satisfy himself why this is so) the quantities f'_A and f'_B defined by the equations

$$f_A^* = f'_A(V_1 + V_2) \quad \text{and} \quad f_B^* = f'_B(V_1 + V_2) \qquad (6.23)$$

are identical with those quantities given in equations (6.18) and (6.19).

The entropy, energy and Helmholtz function of the mixture are therefore given by the equations

$$S_{\text{mixture}} = kN \ln \frac{f_A'(V_1 + V_2)}{N} + kM \ln \frac{f_B'(V_1 + V_2)}{M}$$

$$+ (M + N)k + \frac{E}{T} \tag{6.24}$$

$$E_{\text{mixture}} = kNT^2 \left(\frac{\partial \ln f_A'(V_1 + V_2)}{\partial T} \right)_V + kMT^2 \left(\frac{\partial \ln f_B'(V_1 + V_2)}{\partial T} \right)_V \tag{6.25}$$

and

$$A_{\text{mixture}} = -kNT \ln \frac{f_A'(V_1 + V_2)}{N} - kMT \ln \frac{f_B'(V_1 + V_2)}{M}$$

$$-(N + M)kT. \tag{6.26}$$

6.3   Thermodynamic Quantities of Mixing

It follows from the results of the previous section that the thermodynamic quantities of mixing are obtained by the subtraction of equations (6.20), (6.21) and (6.22) from equations (6.24), (6.25) and (6.26) so that

$$\Delta S_{\text{mixing}} = kN \ln \frac{V_1 + V_2}{V_1} + kM \ln \frac{V_1 + V_2}{V_2}, \tag{6.27}$$

$$\Delta E_{\text{mixing}} = 0, \tag{6.28}$$

and

$$\Delta A_{\text{mixing}} = -kNT \ln \frac{V_1 + V_2}{V_1} - kMT \ln \frac{V_1 + V_2}{V_2}. \tag{6.29}$$

These equations hold whether the original sub-systems were at the same initial pressure or not. Their most valuable feature is that they demonstrate that the non-zero values for ΔS and ΔA are due solely to the fact that the removal of the barrier, so permitting mixing to occur, results in an increase in the volume accessible to the particles and *so to changes in their (transla-tional) energy levels.*

It is useful for some purposes to express these equations in other ways. Suppose first that the sub-systems were at the same initial pressure P. It then follows from (5.75) that

$$V_1 = NkT/P \quad \text{and} \quad V_2 = MkT/P$$

so that

$$\frac{V_1}{V_1 + V_2} = \frac{N}{N + M} = x_A,$$

and

$$\frac{V_2}{V_1 + V_2} = \frac{M}{N + M} = x_B,$$

where x_A and x_B are the *mole fractions* of components A and B in the mixture. In this event equations (6.27) and (6.29) become

$$\Delta S_{mixing} = -kN \ln x_A - kM \ln x_B \qquad (6.30)$$

and

$$\Delta A_{mixing} = kNT \ln x_A + kMT \ln x_B. \qquad (6.31)$$

Suppose now that the sub-systems were originally at different initial pressures P_1 and P_2. We then have

$$V_1 = NkT/P_1 \quad \text{and} \quad V_2 = MkT/P_2$$

and, denoting the final pressure of the mixture by the symbol P_f

$$(V_1 + V_2) = \frac{(N + M)kT}{P_f}.$$

It follows that

$$\Delta S_{mixing} = kN \ln \frac{(N + M)P_1}{NP_f} + kM \ln \frac{(N + M)P_2}{MP_f}$$

$$= kN \ln \frac{P_1}{x_A P_f} + kM \ln \frac{P_2}{x_B P_f}.$$

But $x_A P_f$ is the *partial pressure* of species A in the mixture, and $x_B P_f$ the *partial pressure* of species B, so that denoting these partial pressures by P_A and P_B we find that

$$\Delta S_{mixing} = kN \ln \frac{P_1}{P_A} + kM \ln \frac{P_2}{P_B} \qquad (6.32)$$

and

$$\Delta A_{mixing} = -kNT \ln \frac{P_1}{P_A} - kMT \ln \frac{P_2}{P_B}. \qquad (6.33)$$

6.4   Expressions for the Molecular Chemical Potential of Components of Perfect Gas Mixtures

This is a very important section indeed; the expressions produced here being necessary for the study of chemical equilibrium in the next chapter.

It follows from equation (6.26) that the Helmholtz function for a mixture of N molecules of species A, M molecules of species B . . . contained in volume V is given by the equation

$$A = -kNT \ln \frac{f_A}{N} - kMT \ln \frac{f_B}{M} + \cdots - NkT - MkT + \cdots \quad (6.34)$$

where f_A and f_B are the partition functions for the individual species, and equal of course to the product of the partition functions for each mode of molecular behaviour, so that

$$f_A = (f_t \cdot f_r \cdot f_v \cdot f_e)_A$$

and

$$f_B = (f_t \cdot f_r \cdot f_v \cdot f_e)_B.$$

The volume V does not *appear* in these equations but *determines* the accessible translational energy levels and is *contained* in the translational partition functions f_{t_A} and f_{t_B}.

Since the molecular chemical potentials are defined by the equations

$$\mu_A = \left(\frac{\partial A}{\partial N}\right)_{T,V,M} \quad \text{and} \quad \mu_B = \left(\frac{\partial A}{\partial M}\right)_{T,V,N}$$

it follows from (6.34) that

$$\mu_A = -kT \ln \frac{f_A}{N} \quad (6.35)$$

and

$$\mu_B = -kT \ln \frac{f_B}{M}. \quad (6.36)$$

These equations can be put into different forms. We shall develop alternative forms only for μ_A since those for μ_B are of course analogous.

We first express f_A as $(f'_t V \cdot f_{int})_A$ where $f'_t = f_t/V$ and $f_{int} = f_r \cdot f_v \cdot f_e$, and then replace V/N by kT/P_A where P_A is obviously the pressure exerted by N molecules of A confined to volume V at temperature T, (so that P_A is the partial pressure of A in the mixture), and so obtain the equation

$$\mu_A = -kT \ln \frac{f'_t \cdot f_{int} kT}{P_A}. \tag{6.37}$$

We now multiply[1] top and bottom of the argument by $P^\dagger$ where $P^\dagger$ is a chosen 'standard' pressure and then split the logarithmic term to give

$$\mu_A = -kT \ln \frac{f'_t \cdot f_{int} \cdot kT}{P^\dagger} + kT \ln \frac{P_A}{P^\dagger} \tag{6.38}$$

i.e.

$$\mu_A = \mu_A^\dagger + kT \ln \frac{P_A}{P^\dagger} \tag{6.39}$$

where $\mu_A^\dagger$, the so-called 'standard' chemical potential of A, i.e. the value of the chemical potential at the chosen standard pressure $P^\dagger$, is given by the equation

$$\mu_A^\dagger = -kT \ln \frac{f'_t \cdot f_{int} \cdot kT}{P^\dagger}. \tag{6.40}$$

In the case of a monatomic gas at temperatures at which electronic excitation is negligible,[2]

$$\mu_A^\dagger = -kT \ln \frac{(2\pi m_A kT)^{\frac{3}{2}} kT g_{e_0}}{h^3 P^\dagger} \tag{6.41}$$

the chosen energy zero being that of zero translational energy and that of the ground electronic state. Comparison of this equation with (5.95) shows that the quantity $\mu_A^\dagger$ which figures in the mechanics of perfect gas *mixtures* is precisely the same as the corresponding quantity figuring in the mechanics of a single component gas. In other words the quantity $\mu_A^\dagger$ figuring in the above equations is the chemical potential of A as a single component gas at pressure $P^\dagger$. If we insert the value $P^\dagger = 1 \text{ Nm}^{-2}$ into (6.41) we obtain the value for $\mu_A^\dagger$ at that pressure; if we insert the value $P^\dagger = 101.325 \text{ kNm}^{-2}$ into (6.41) we obtain the value for $\mu_A^\dagger$ at a pressure of one atmosphere.

Corresponding expressions for the chemical potential of diatomic (and polyatomic) molecules are slightly more complicated, first because of the necessary inclusion of rotational and vibrational terms, and secondly because we have a *choice* of energy zeros.

---

[1] This step ensures that when we split the logarithmic term the argument of *each* logarithmic term is dimensionless as it must be.

[2] If the temperature is such that electronic excitation is other than negligible the term g_{e_0} must of course be replaced by the term $g_{e_0} + g_{e_1} \exp(-\Delta\epsilon_{e_1}/kT) + \cdots$.

It follows from (6.40) and the equations of Chapter 5, that the standard chemical potential for a diatomic gas at temperatures at which it is both rotationally[1] and vibrationally excited but at which electronic excitation is negligible, is given by the expression

$$\mu_A^\dagger = -kT \ln \left[\frac{(2\pi m kT)^{\frac{3}{2}} kT}{h^3 P^\dagger} \cdot \frac{T}{\sigma\theta_r} \cdot \frac{g_{e_0}}{1 - \exp(-\theta_v/T)} \right] \qquad (6.42)$$

the chosen energy zero being that of the hypothetical state corresponding to zero translational and rotational activity but to that of the *ground* vibrational and electronic states, or alternatively by the expression

$$\mu_A^\dagger = -kT \ln \left[\frac{(2\pi m kT)^{\frac{3}{2}} kT}{h^3 P^\dagger} \cdot \frac{T}{\sigma\theta_r} \cdot \frac{g_{e_0}}{1 - \exp(-\theta_v/T)} \right] - D_0 \qquad (6.43)$$

the chosen energy zero being now that of the hypothetical state corresponding to zero translational and rotational energy and to the ground electronic state but that in which the ground state vibrational energy is measured relative to the potential energy of the constituent atoms at infinite separation, D_0 being the molecular dissociation energy at absolute zero. (This second convention is the one we shall require to adopt when, in the next chapter we use expressions for $\mu^\dagger$ in the calculation of equilibrium constants.) Comparison of (6.42) and 6.43) with (5.98) and (5.99) shows again that the quantity $\mu^\dagger$ figuring here is precisely the same as the corresponding quantity figuring in the mechanics of a single component gas.

Expressions for the standard chemical potential of a linear polyatomic molecule containing n atoms differ from equations (6.42) and (6.43) only in so far that the vibrational term $[1 - \exp(-\theta_v/T)]$ is replaced by the product of $3n - 5$ such terms as explained in Section 5.3, whilst corresponding expressions in the case of a non-linear molecule differ only in so far that the rotational term $T/\sigma\theta_r$ is replaced by the term $\pi^{\frac{1}{2}}/\sigma[T^3/\theta_1\theta_2\theta_3]^{\frac{1}{2}}$, and the vibrational term by the product of $3n - 6$ such terms. Thus the corresponding expression to (6.43) for a linear triatomic molecule (characterised by four vibrational modes) is

$$\mu^\dagger = -kT \ln \left[\frac{(2\pi m kT)^{\frac{3}{2}} kT}{h^3 P^\dagger} \cdot \frac{T}{\sigma\theta_r} \cdot \frac{g_{e_0}}{\prod_{1,2,3,4}[1 - \exp(-\theta_v/T)]} \right] - D_0 \qquad (6.44)$$

whilst that for a non-linear triatomic molecule (characterised by three

---

[1] If T is not very much greater than θ_r, the term T/θ_r must of course be replaced by the expression given in equation (5.23).

vibrational modes) is

$$\mu^\dagger = -kT \ln \left[\frac{(2\pi mkT)^{\frac{3}{2}}kT}{h^3 P^\dagger} \cdot \frac{\pi^{\frac{1}{2}}}{\sigma}\left(\frac{T^3}{\theta_1\theta_2\theta_3}\right)^{\frac{1}{2}} \cdot \frac{g_{e_0}}{\prod_{1,2,3}[1 - \exp{(-\theta_v/T)}]} \right] - D_0.$$

(6.45)

In some instances it is convenient to use expressions for the chemical potential of a component of a gas mixture in terms of its mole fraction. These are obtained easily from the equations given above. The mole fraction of a component A in a mixture containing N molecules of A, M molecules of B ... is defined by the equation

$$x_A = \frac{N}{N + M + \cdots},$$

and it follows that the mole fraction and partial pressure of a component are related by the equation

$$P_A = x_A P$$

where P is the total pressure of the mixture. Substitution into (6.38) leads therefore to the equation

$$\mu_A = -kT \ln \frac{f_t' \cdot f_{int} \cdot kT}{P} + kT \ln x_A$$

$$= \mu_A^\circ + kT \ln x_A$$

(6.46)

where μ_A° is the chemical potential of A as a single component gas at pressure P. Expressions for μ_A° are of course obtained from equations (6.41) to (6.45) simply by replacing $P^\dagger$ by P.

The reader is reminded that the chemical potential (partial *molar* free energy) occurring in classical thermodynamics, and in this book denoted by the symbol μ^* is related to the molecular chemical potential by the equation $L\mu = \mu^*$. It follows from (6.39), (6.40) and (6.46) that

$$\mu_A^* = \mu_A^{\dagger*} + RT \ln \frac{P_A}{P^\dagger}$$

(6.47)

where

$$\mu_A^{\dagger*} = -RT \ln \frac{f_t' \cdot f_{int} \cdot kT}{P^\dagger}$$

(6.48)

and

$$\mu_A^* = \mu_A^{\circ*} + RT \ln x_A$$

(6.49)

where

$$\mu_A^{\circ}* = -RT\ln\frac{f_t' \cdot f_{\text{int}} \cdot kT}{P}. \qquad (6.50)$$

The standard pressure almost always adopted in classical thermodynamics is that of one atmosphere. The evaluation of $\mu_A^{\dagger}*$ requires therefore the use of the value $P^{\dagger} = 101.325\,\text{kNm}^{-2}$ in (6.48) and that the pressure term in (6.47) is expressed in the same units.

CHAPTER 7

The Principles of Chemical Equilibrium in Perfect Gas Reactions

7.1 The Statistical–Mechanical Criterion for Chemical Equilibrium

Consider an isolated system of total energy E and volume V consisting of a mixture of perfect gases A, B, C, D, ... and suppose that the system contains N_A^* molecules of A, N_B^* molecules of B, N_C^* of C and N_D^* of D, ... If no chemical reaction is possible (so that the numbers N_A^*, N_B^*,... remain unchanged) the number of complexions corresponding to the most probable distribution among energy states is given, as shown in Section 6.2 by the equation

$$\ln W^* = N_A^* \ln \frac{f_A^*}{N_A^*} + N_A^* + N_B^* \ln \frac{f_B^*}{N_B^*} + N_B^*$$

$$+ N_C^* \ln \frac{f_C^*}{N_C^*} + N_C^* + N_D^* \ln \frac{f_D^*}{N_D^*} + N_D^* + \cdots + \frac{E}{kT^*}, \qquad (7.1)$$

where T^* is the temperature and $f_A^*, f_B^*, \ldots$ the corresponding molecular partition functions, each partition function being of course defined by an equation

$$f_A^* = \sum_i g_{A_i} \exp\left(-\epsilon_i/kT^*\right).$$

Let us now suppose that a chemical reaction

$$a\mathrm{A} + b\mathrm{B} + \cdots \rightleftharpoons c\mathrm{C} + d\mathrm{D}$$

occurs. What changes in the quantities $W^*, f_A^*, f_B^*, \ldots$ and T^* result? The energy levels accessible to each molecule are determined solely by the volume

V and so are unchanged as are the degeneracies of each level, but because chemical reaction disturbs the original division of the total energy E between that associated with the chemical bonds and that associated with the thermal activity of the molecules, the temperature of the system is changed as are therefore the values of the partition functions, and since the reaction is spontaneous the W value must presumably increase. It is evident however that to each resultant composition there corresponds an equation analogous completely to (7.1) and that the composition of the system when chemical equilibrium is reached is determined by those values of $N_A, N_B, N_C, N_D, \ldots$ which maximise the value of W where

$$\ln W = N_A \ln \frac{f_A}{N_A} + N_A + N_B \ln \frac{f_B}{N_B} + N_B + N_C \ln \frac{f_C}{N_C} + N_C$$

$$+ N_D \ln \frac{f_D}{N_D} + N_D + \cdots + \frac{E}{kT} \tag{7.2}$$

and where $f_A, f_B, f_C, f_D, \ldots$ are the values of the partition functions at the final equilibrium temperature T.

It remains only to show how these values of N_A, N_B, N_C, N_D may be determined.

If we suppose that small amounts of A and B ... react to form C and D ... we may define a quantity $d\xi$ in terms of the increase in the number of molecules of each species by the equations

$$d\xi = -\frac{dN_A}{a} = -\frac{dN_B}{b} = \frac{dN_C}{c} = \frac{dN_D}{d} = \cdots \tag{7.3}$$

the quantity ξ being the de Donder extent of reaction.

It follows that the equilibrium values of $N_A, N_B, \ldots$ are those given by the equation

$$\left(\frac{\partial \ln W}{\partial \xi} \right)_{E,V} = 0 \tag{7.4}$$

where $\ln W$ is given by equation (7.2).

Making use of (7.3) we see that

$$\left(\frac{\partial \ln W}{\partial \xi} \right)_{E,V} = 0 = -a \left[\frac{\partial \{N_A \ln (f_A/N_A) + N_A\}}{\partial N_A} \right]_{E,V}$$

$$- b \left[\frac{\partial \{N_B \ln (f_B/N_B) + N_B\}}{\partial N_B} \right]_{E,V}$$

$$+ c \left[\frac{\partial \{N_C \ln (f_C/N_C) + N_C\}}{\partial N_C} \right]_{E,V}$$

$$+ d\left[\frac{\partial\{N_D \ln(f_D/N_D) + N_D\}}{\partial N_D}\right]_{E,V}$$

$$+ \frac{E}{k}\left(\frac{\partial(1/T)}{\partial\xi}\right)_{E,V}$$

$$= - a\ln\frac{f_A}{N_A} - b\ln\frac{f_B}{N_B} + c\ln\frac{f_C}{N_C} + d\ln\frac{f_D}{N_D}$$

$$- \left[aN_A\frac{\partial\ln f_A}{\partial N_A} + bN_B\frac{\partial\ln f_B}{\partial N_B} - cN_C\frac{\partial\ln f_C}{\partial N_C}\right.$$

$$\left. - dN_D\frac{\partial\ln f_D}{\partial N_D} - \frac{E}{k}\frac{\partial(1/T)}{\partial\xi}\right]_{E,V}. \tag{7.5}$$

The disposal of the quantities contained in the last brackets is an exercise similar to ones we have studied before; (see in particular the derivation of equation 2.24).

$$aN_A\frac{\partial\ln f_A}{\partial N_A} = a\frac{N_A}{f_A}\cdot\frac{\partial(1/T)}{\partial N_A}\cdot\frac{\partial f_A}{\partial(1/T)}$$

$$= -\frac{a}{k}\cdot\frac{\partial(1/T)}{\partial N_A}\cdot\sum_i n_{A_i}\epsilon_{A_i}$$

$$= \frac{1}{k}\cdot\frac{\partial(1/T)}{\partial\xi}\cdot\sum_i n_{A_i}\epsilon_{A_i}. \tag{7.6}$$

Similar equations may be obtained for the analogous terms relating to species B, C, D, ... so that

$$aN_A\frac{\partial\ln f_A}{\partial N_A} + bN_B\frac{\partial\ln f_B}{\partial N_B} - cN_C\frac{\partial\ln f_C}{\partial N_C} - dN_D\frac{\partial\ln f_D}{\partial N_D} - \frac{E}{k}\frac{\partial(1/T)}{\partial\xi}$$

$$= \frac{1}{k}\left[\sum_i n_{A_i}\epsilon_{A_i} + \sum_j n_{B_j}\epsilon_{B_j} + \sum_k n_{C_k}\epsilon_{C_k} + \sum_l n_{D_l}\epsilon_{D_l} - E\right]\frac{\partial(1/T)}{\partial\xi}. \tag{7.7}$$

The quantity in brackets in this equation is obviously zero since the sum of the first four terms is the energy of the system, so that (7.5) reduces to the form

$$\left(\frac{\partial\ln W}{\partial\xi}\right)_{E,V} = 0$$

$$= - a\ln\frac{f_A}{N_A} - b\ln\frac{f_B}{N_B} + c\ln\frac{f_C}{N_C} + d\ln\frac{f_D}{N_D}. \tag{7.8}$$

We conclude that if in an isolated mixture of perfect gases A, B, C, D, ... a chemical reaction

$$aA + bB + \cdots \rightleftharpoons cC + dD + \cdots$$

is possible, reaction proceeds until the numbers of molecules of each species reach such values $N_A, N_B, N_C, N_D, \ldots$ that

$$-a \ln \frac{f_A}{N_A} - b \ln \frac{f_B}{N_B} - \cdots = -c \ln \frac{f_C}{N_C} - d \ln \frac{f_D}{N_D} - \cdots . \qquad (7.9)$$

This is the fundamental criterion for chemical equilibrium. It can be put into a rather more familiar form by making use of the relationships established in Section 6.4. There it was shown that the chemical potentials of perfect gases A, B, ... in a mixture containing N_A molecules of A, N_B molecules of B, ... are given by the equations

$$\mu_A = -kT \ln \frac{f_A}{N_A}, \qquad \mu_B = -kT \ln \frac{f_B}{N_B}$$

and so on. It follows that the criterion for chemical equilibrium for the reaction

$$aA + bB + \cdots \rightleftharpoons cC + dD + \cdots$$

may be written as

$$a\mu_{A_{eq}} + b\mu_{B_{eq}} + \cdots = c\mu_{C_{eq}} + d\mu_{D_{eq}} + \cdots . \qquad (7.10)$$

7.2 Energy Zeros

Equation (7.10) obviously has meaning only if the chemical potentials of all species are measured from the same energy zero. This is just the situation envisaged in Section 6.4, and as explained there may be achieved by measuring the chemical potential of the molecule against that of the completely separated atoms at rest, (i.e. a hypothetical state in which the translational and rotational energies are zero), each atom is in its electronic ground state and the ground state vibrational energy of the molecule is measured against the potential energy of the constituent atoms at infinite separation. The appropriate equation for the standard chemical potentials of diatomic molecules is therefore (6.43), that for linear polyatomic molecules (6.44) and that for non-linear polyatomic molecules (6.45).

7.3 Equilibrium Constants

The equilibrium condition

$$a\mu_{A_{eq}} + b\mu_{B_{eq}} + \cdots = c\mu_{C_{eq}} + d\mu_{D_{eq}} \qquad (7.10)$$

for the reaction

$$aA + bB + \cdots \rightleftharpoons cC + dD + \cdots$$

leads directly to expressions for the so-called equilibrium constants for the reaction. Two lines of approach are possible and we shall explore each in turn.

Approach 1

It follows from (6.35) that

$$\mu_{A_{eq}} = -kT \ln \frac{f_A}{N_{A_{eq}}}$$

and that similar equations may be written for $\mu_{B_{eq}}$, $\mu_{C_{eq}}$ and $\mu_{D_{eq}}$. Substituting into (7.10) and rearranging, we see immediately that

$$\frac{N_{C_{eq}}^c N_{D_{eq}}^d}{N_{A_{eq}}^a N_{B_{eq}}^b} = \frac{f_C^c f_D^d}{f_A^a f_B^b}. \tag{7.11}$$

All terms on the right-hand side are functions of the final equilibrium temperature and (through the translational partition functions) of the volume of the system, but are independent of the amounts of each species. We can therefore write

$$\frac{N_{C_{eq}}^c N_{D_{eq}}^d}{N_{A_{eq}}^a N_{B_{eq}}^b} = K_N(T, V), \tag{7.12}$$

the quantity K_N being known as the equilibrium constant for the reaction in terms of the numbers of molecules of the components at equilibrium. We observe that since for each component

$$f = f'_t \cdot V \cdot f_{\text{int}}$$

where

$$f'_t = \frac{(2\pi m k T)^{\frac{3}{2}}}{h^3} \quad \text{and} \quad f_{\text{int}} = f_r \cdot f_v \cdot f_e,$$

K_N is independent of the volume of the system only if the stoichiometric coefficients $a, b \ldots c, d \ldots$ are such that $c + d \ldots = a + b \ldots$.

Alternatively we may choose to express the chemical potentials of the components by equation (6.38), i.e.

$$\mu_A = -kT \ln \frac{f'_t \cdot f_{\text{int}} \cdot kTP^\dagger}{P^\dagger P_{A_{eq}}},$$

whence we find that

$$\frac{(P_{C_{eq}}/P^\dagger)^c(P_{D_{eq}}/P^\dagger)^d}{(P_{A_{eq}}/P^\dagger)^a(P_{B_{eq}}/P^\dagger)^b} = \frac{\left[\frac{(f'_t f_{int})_C kT}{P^\dagger}\right]^c\left[\frac{(f'_t f_{int})_D kT}{P^\dagger}\right]^d}{\left[\frac{(f'_t f_{int})_A kT}{P^\dagger}\right]^a\left[\frac{(f'_t f_{int})_B kT}{P^\dagger}\right]^b}. \tag{7.13}$$

All terms on the right-hand side are functions of temperature only (although the numerical value of the term in each bracket depends on the chosen standard pressure $P^\dagger$). We can therefore write

$$\frac{(P_{C_{eq}}/P^\dagger)^c(P_{D_{eq}}/P^\dagger)^d}{(P_{A_{eq}}/P^\dagger)^a(P_{B_{eq}}/P^\dagger)^b} = K_{P/P^\dagger}(T) \tag{7.14}$$

where $K_{P/P^\dagger}$ is known as the equilibrium constant for the reaction in terms of the equilibrium partial pressures.

Alternatively we may choose to express the chemical potentials of the components by equation (6.46) i.e.

$$\mu_{A_{eq}} = -kT \ln \frac{f'_t f_{int} kT}{x_{A_{eq}} P}$$

where $x_{A_{eq}}$ is the mole fraction of component A at equilibrium and P is the total pressure of the system at equilibrium. Substituting into (7.10) and rearranging we find that

$$\frac{x_{C_{eq}}^c x_{D_{eq}}^d}{x_{A_{eq}}^a x_{B_{eq}}^b} = \frac{\left[\frac{(f'_t f_{int})_C kT}{P}\right]^c\left[\frac{(f'_t f_{int})_D kT}{P}\right]^d}{\left[\frac{(f'_t f_{int})_A kT}{P}\right]^a\left[\frac{(f'_t f_{int})_B kT}{P}\right]^b}. \tag{7.15}$$

All terms on the right-hand side are functions of temperature and of the pressure at equilibrium. We therefore write

$$\frac{x_{C_{eq}}^c x_{D_{eq}}^d}{x_{A_{eq}}^a x_{B_{eq}}^b} = K_x(T, P) \tag{7.16}$$

the quantity K_x being known as the equilibrium constant for the reaction in terms of the equilibrium mole fractions. We observe that K_x is independent of P only if $c + d \cdots = a + b \cdots$, and the numerical value of K_x independent of the units in which P is expressed only in the same circumstances.

The approach described above is attractive because of its inherent simplicity, but the alternative approach is perhaps more instructive in some ways, and to this we now turn. We consider only the alternative approach to $K_{P/P^\dagger}$ and K_x.

Approach 2

It follows from equation (6.39) that

$$\mu_{A_{eq}} = \mu_A^\dagger + kT \ln \frac{P_{A_{eq}}}{P^\dagger}$$

where $\mu_A^\dagger$ is the value of the (molecular) chemical potential of A at the chosen standard pressure $P^\dagger$. Similar expressions may be written for the other components. Substituting these into (7.10) and rearranging we find that

$$kT \ln \frac{(P_{C_{eq}}/P^\dagger)^c(P_{D_{eq}}/P^\dagger)^d}{(P_{A_{eq}}/P^\dagger)^a(P_{B_{eq}}/P^\dagger)^b} = a\mu_A^\dagger + b\mu_B^\dagger \ldots - c\mu_C^\dagger - d\mu_D^\dagger \ldots . \quad (7.17)$$

Each of the quantities $\mu_A^\dagger, \mu_B^\dagger, \ldots$ is a function only of temperature, and in particular is independent of the number of molecules of each species present. It follows that the quantity

$$\frac{(P_{C_{eq}}/P^\dagger)^c(P_{D_{eq}}/P^\dagger)^d}{(P_{A_{eq}}/P^\dagger)^a(P_{B_{eq}}/P^\dagger)^b}$$

is a function only of temperature. As shown earlier it is given the symbol $K_{P/P^\dagger}$. We may therefore write

$$kT \ln K_{P/P^\dagger} = a\mu_A^\dagger + b\mu_B^\dagger \cdots - c\mu_C^\dagger - d\mu_D^\dagger \cdots . \quad (7.18)$$

It will be recalled that each of the standard molecular chemical potentials $\mu_A^\dagger, \mu_B^\dagger, \ldots$ is related to the corresponding standard *molar* chemical potential used in classical thermodynamics by the equation $L\mu_A^\dagger = \mu_A^{*\dagger}$ where L is the Avogadro constant, and since $kL = R$, the molar gas constant, we find by multiplying both sides of (7.18) by L that

$$RT \ln K_{P/P^\dagger} = a\mu_A^{*\dagger} + b\mu_B^{*\dagger} \cdots - c\mu_C^{*\dagger} - d\mu_D^{*\dagger} \cdots . \quad (7.19)$$

The right-hand side of this equation represents the decrease in Gibbs function for the process in which a moles of a pure gas A at the standard pressure $P^\dagger$, and b moles of a pure gas B at the same pressure react to form c moles of a pure gas C and d moles of a pure gas D each at the same pressure $P^\dagger$. This quantity is denoted by the symbol $-\Delta G^\dagger$. We reach therefore the equation

$$RT \ln K_{P/P^\dagger} = -\Delta G^\dagger \quad (7.20)$$

which is of course one of the best known and most widely used relations of classical thermodynamics.

We arrive at analogous expressions for K_x by making use of equation (6.46), so that

$$\mu_{A_{eq}} = \mu_A^\circ + kT \ln x_{A_{eq}}$$

where μ_A° is the (molecular) chemical potential of A as a single-component gas at pressure P. Substituting this and similar expressions for $\mu_{B_{eq}}, \dots$ into (7.10) we find that

$$kT \ln \frac{x_{C_{eq}}^c x_{D_{eq}}^d}{x_{A_{eq}}^a x_{B_{eq}}^b} = a\mu_A^\circ + b\mu_B^\circ \cdots - c\mu_C^\circ - d\mu_D^\circ \cdots . \qquad (7.21)$$

It follows that the quantity $x_{C_{eq}}^c x_{D_{eq}}^d / x_{A_{eq}}^a x_{B_{eq}}^b$ is a function only of T and P. It is, as shown earlier given the symbol K_x, so that

$$kT \ln K_x = a\mu_A^\circ + b\mu_B^\circ \cdots - c\mu_C^\circ - d\mu_D^\circ \cdots . \qquad (7.22)$$

Multiplying both sides by L we see that

$$RT \ln K_x = a\mu_A^{*\circ} + b\mu_B^{*\circ} \cdots - c\mu_C^{*\circ} - d\mu_D^{*\circ} \cdots \qquad (7.23)$$

where each of the terms $\mu^{*\circ}$ is the corresponding standard molar quantity. The right-hand side of this equation represents the decrease in Gibbs function occasioned by the conversion of a moles of a pure gas A and b moles of a pure gas B each at pressure P to c moles of C and d moles of D each at the same pressure P. This quantity is given the symbol $-\Delta G^\circ$. We reach therefore the equation

$$RT \ln K_x = -\Delta G^\circ. \qquad (7.24)$$

7.4 The Evaluation of Equilibrium Constants from Statistical–Mechanical Data

We now turn to the evaluation of $K_{P/P^\dagger}$ and K_x from statistical–mechanical data. The starting point for the first is equation (7.18) but because the detailed content of the expression for each depends on the nature of the molecules concerned we choose a definite reaction.

Consider the reaction

$$H_2 + \tfrac{1}{2}O_2 \rightleftharpoons H_2O$$

at 1000 K, at which temperature we may correctly suppose (a) that the rotational activity of all three species may be represented by the "high temperature formulae" of Section 5.2, (b) that all molecules are vibrationally excited, but (c) that all molecules remain in their ground electronic state. It follows, bearing in mind the provisions of Section 7.2 that the appropriate expressions for $\mu_{(H_2)}^\dagger$ and $\mu_{(O_2)}^\dagger$ are those obtained from equation (6.43) whilst that for $\mu_{(H_2O)}^\dagger$ (H_2O being a triatomic non-linear molecule characterised by three vibrational modes) is obtained from equation (6.45), i.e.

$$\mu_{(H_2)}^\dagger = -kT\ln\left[\frac{(2\pi m_{(H_2)}kT)^{\frac{3}{2}}kT}{h^3 P^\dagger}\cdot\frac{T}{(\sigma\theta_r)_{(H_2)}}\cdot\frac{g_{e0(H_2)}}{1-\exp(-\theta_v/T)}\right] - D_{0(H_2)}.$$

$$\mu_{(O_2)}^\dagger = -kT\ln\left[\frac{(2\pi m_{(O_2)}kT)^{\frac{3}{2}}kT}{h^3 P^\dagger}\cdot\frac{T}{(\sigma\theta_r)_{(O_2)}}\cdot\frac{g_{e0(O_2)}}{1-\exp(-\theta_v/T)}\right] - D_{0(O_2)}$$

and

$$\mu_{(H_2O)}^\dagger = -kT\ln\left[\frac{(2\pi m_{(H_2O)}kT)^{\frac{3}{2}}kT}{h^3 P^\dagger}\cdot\frac{1}{\sigma_{(H_2O)}}\left(\frac{\pi T^3}{\theta_1\theta_2\theta_3}\right)^{\frac{1}{2}}\cdot\right.$$
$$\left.\frac{g_{e0(H_2O)}}{\prod_{1,2,3}[1-\exp(-\theta_v/T)]}\right] - D_{0(H_2O)}.$$

Substituting into the equation

$$kT\ln K_{P/P^\dagger} = \mu_{H_2}^\dagger + \tfrac{1}{2}\mu_{O_2}^\dagger - \mu_{H_2O}^\dagger$$

and rearranging we find that

$$K_{P/P^\dagger} = \frac{\left[\dfrac{(2\pi m_{(H_2O)}kT)^{\frac{3}{2}}kT}{h^3 P^\dagger}\right]}{\left[\dfrac{(2\pi m_{(H_2)}kT)^{\frac{3}{2}}kT}{h^3 P^\dagger}\right]\left[\dfrac{(2\pi m_{(O_2)}kT)^{\frac{3}{2}}kT}{h^3 P^\dagger}\right]^{\frac{1}{2}}}$$

$$\times\frac{\dfrac{1}{\sigma_{(H_2O)}}\left[\dfrac{\pi T^3}{\theta_1\theta_2\theta_3}\right]^{\frac{1}{2}}}{\left[\dfrac{T}{(\sigma\theta_r)_{H_2}}\right]\left[\dfrac{T}{(\sigma\theta_r)_{O_2}}\right]^{\frac{1}{2}}}$$

$$\times\frac{[1-\exp(-\theta_{v(H_2)}/T)][1-\exp(-\theta_{v(O_2)}/T)]^{\frac{1}{2}}}{\prod_{1,2,3}[1-\exp(-\theta_{v(H_2O)}/T)]}$$

$$\times\frac{g_{e0(H_2O)}}{g_{e0(H_2)}[g_{e0(O_2)}]^{\frac{1}{2}}}$$

$$\times\exp\left[\frac{D_{0(H_2O)}-D_{0(H_2)}-\frac{1}{2}D_{0(O_2)}}{kT}\right]. \tag{7.25}$$

We turn now to the evaluation of K_x for the same reaction. The starting point is now equation (7.22) but we need spend little time on the exercise because as pointed out in Section 6.4, expressions for $\mu°$ are obtained from those for $\mu^\dagger$ simply by replacing $P^\dagger$ by P, so that the expression for K_x for the reaction

$H_2 + \frac{1}{2}O_2 \rightleftharpoons H_2O$ is obtained from (7.25) simply by making this substitution. We take this opportunity however of repeating that although $K_{P/P\dagger}$ is independent of the actual pressure of the system the value of K_x is so dependent, and since the partial pressure of a gas is connected with its mole fraction by the equation $P_A = x_A P$ where P is the total pressure,

$$K_{P/P\dagger} = K_x \left(\frac{P}{P\dagger} \right)^{(c+d\cdots-a-b\cdots)} \tag{7.26}$$

We shall not attempt to gather together (or cancel) like terms in equation (7.25) because the expression serves more usefully as an example if left in its present state. We remark only that supposing we wish to evaluate the expression in terms of S.I. units, m, the mass of each molecule must be expressed in kilo-grammes and $P\dagger$ must be expressed in Newtons per square metre. The numerical value for $K_{P/P\dagger}$ depends on the chosen standard pressure (unless for the reaction concerned $c + d \cdots = a + b \cdots$) and by substituting $P\dagger = 1 \, \text{Nm}^{-2}$ into (7.25) we obtain one value for $K_{P/P\dagger}$ but if instead we choose to adopt the standard pressure of one atmosphere we must insert the value $P\dagger = 101.325 \, \text{kNm}^{-2}$ into (7.25) which will of course give a different value for $K_{P/P\dagger}$. (The same remarks apply of course to the evaluation of K_x, the numerical value obtained depends on the unit in which the total pressure P is expressed.)

One further point should be made. Although the numerical value of $K_{P/P\dagger}$ depends on the chosen standard pressure $P\dagger$, $K_{P/P\dagger}$ is itself a dimensionless quantity. In many chemistry texts equilibrium constants are often the schematic representation

$$K_P = \frac{P_{C_{eq}}^c P_{D_{eq}}^d}{P_{A_{eq}}^a P_{B_{eq}}^b}$$

so that (except for reactions in which $c + d \cdots = a + b \cdots$) K_P has the dimensions of pressure raised to the power $(c + d \cdots - a - b \cdots)$. There is of course no objection to the use of K_P instead of its dimensionless analogue $K_{P/P\dagger}$ and as will be seen in a later section we shall report the value for the equilibrium constant for the reaction

$$Cs \rightleftharpoons Cs + \bar{e}$$

as

$$K_{P/P\dagger} = 4.72 \quad \text{if } P\dagger = 1 \, \text{Nm}^{-2}$$

or as

$$K_{P/P\dagger} = 4.65 \times 10^{-5} \quad \text{if } P\dagger = 1 \, \text{atm} = 101.325 \, \text{kNm}^{-2}$$

or as

$$K_P = 4.72 \text{ Nm}^{-2} = 4.65 \times 10^{-5} \text{ atm},$$

but a dimensionless equilibrium constant *must* of course be used if we wish to write ln K (as we have done in equation 7.18) since the logarithm of an argument which is not a pure number is meaningless.

7.5 The Meaning and Evaluation of the Exponential Term

We must now consider the significance of the term

$$\exp \frac{D_{0(H_2O)} - D_{0(H_2)} - \frac{1}{2}D_{0(O_2)}}{kT}$$

figuring in equation (7.25). Since D_0 is the difference in energy between the lowest vibrational level and the atoms separated at infinity, it is clear from the schematic representation

$$H_2 \xrightarrow{D_{0(H_2)}} 2H \qquad \begin{array}{c} H \\ \diagdown \\ \xrightarrow{-D_{0(H_2O)}} \quad O \\ \diagup \\ H \end{array}$$

$$\tfrac{1}{2}O_2 \xrightarrow{\frac{1}{2}D_{0(O_2)}} O$$

that $L[D_{0(H_2)} + \frac{1}{2}D_{0(O_2)} - D_{0(H_2O)}]$ is a measure of the energy change at absolute zero for the reaction in which one mole of H_2O is formed from its constituent elements H_2 and O_2. The exponential term in (7.25) can therefore be written as $\exp -(\Delta E^0/RT)$ where

$$\Delta E^0 = E^0(H_2O) - E^0(H_2) - \tfrac{1}{2}E^0(O_2). \tag{7.27}$$

We now turn to the evaluation of ΔE^0. In some cases values for the dissociation energies D_0 for each molecular species may be obtained from spectroscopic readings, in which case ΔE^0 for the reaction

$$aA + bB \rightarrow cC$$

is given by the equation

$$\Delta E^0 = L[aD_{0(A)} + bD_{0(B)} - cD_{0(C)}], \tag{7.28}$$

but more usually and conveniently the value for ΔE^0 is obtained from the published values of the enthalpy functions $(H^T - E^0)/T$ discussed in Section 5.9, and the values for the molar heats of formation of each species at any one temperature.

Let us consider the calculation of ΔE^0 for the reaction

$$CO + H_2O \rightarrow CO_2 + H_2.$$

We have available the following data in Table 5.8:

(a)

$$\Delta H_f^{298.15}(CO) = -110.52 \text{ kJ mol}^{-1},$$
$$\Delta H_f^{298.15}(CO_2) = -393.51 \text{ kJ mol}^{-1},$$
$$\Delta H_f^{298.51}(H_2O) = -241.83 \text{ kJ mol}^{-1}.$$

ΔH_f for H_2 is zero by definition so that the heat of reaction at 298.15 K is

$$\Delta H^{298.15} = -393.51 + 110.52 + 241.83$$
$$= -41.16 \text{ kJ}.$$

(b) For $T = 298.15$ K

$$\frac{H - E^0}{T}(CO) = 29.09 \text{ JK}^{-1} \text{ mol}^{-1},$$

$$\frac{H - E^0}{T}(CO_2) = 31.41 \text{ JK}^{-1} \text{ mol}^{-1},$$

$$\frac{H - E^0}{T}(H_2O) = 33.20 \text{ JK}^{-1} \text{ mol}^{-1},$$

and

$$\frac{H - E^0}{T}(H_2) = 28.40 \text{ JK}^{-1} \text{ mol}^{-1}.$$

It follows that

$$\Delta E^0 = \Delta H^{298.15} - T[31.41 + 28.40 - 33.20 - 29.09]$$
$$= -41\,160 - 298.15(-2.48)$$
$$= -40\,422 \text{ J}.$$

It should perhaps be mentioned that values for heats of formation of compounds (and those for the energies D_0) are usually known with rather less certainty than are the values for the other parameters ($m, \theta_r, \theta_v, \ldots$) required in the evaluation of equations analogous to (7.25), and any uncertainty in the value of an equilibrium constant calculated from statistical data usually stems from errors in the exponential term.

7.6 The Evaluation of $K_{P/P\dagger}$ for Two Reactions of Interest

The reaction $H_2 + D_2 \rightleftharpoons 2HD$

The value of $K_{P/P\dagger}$ for this reaction at 500 K is calculated to illustrate the fact that the somewhat formidable arithmetic encountered in evaluating an expression such as that in (7.25) almost completely disappears (a) if the stoichiometric numbers are such that $a + b \cdots = c + d \cdots$ so that all terms in the translational partition functions except the molecular masses cancel, (b) if the number of linear molecules on both sides of the chemical equation are the same and the number of non-linear molecules the same so that the rotational term is simplified, and (c) if the molecules are all in their vibrational ground state (i.e. if $\theta_v \gg T$ in all cases.) Thus for the reaction $H_2 + D_2 \rightleftharpoons 2HD$ at 500 K the general equation reduces to

$$K_{P/P\dagger} = \left[\frac{M_{(HD)}^2}{M_{(H_2)}M_{(D_2)}} \right]^{\frac{3}{2}} \cdot \frac{\sigma_{(H_2)}\sigma_{(D_2)}}{[\sigma_{(HD)}]^2} \cdot \frac{\theta_{r(H_2)}\theta_{r(D_2)}}{[\theta_{r(HD)}]^2} \cdot \exp\left(-\frac{\Delta E^0}{RT} \right)$$

where here we use M to denote the conventional dimensionless molecular weights, and have written the product of the symmetry numbers separately for a reason which will be apparent later.

We have the following data:

	M	σ	θ_r/K	D_0/eV
H_2	2.015	2	85.4	4.476
D_2	4.028	2	42.7	4.553
HD	3.022	1	64.0	4.511

Consider first the evaluation of the quantity ΔE^0.

$$D_{0(H_2)} + D_{0(D_2)} - 2D_{0(HD)} = 0.007 \text{ eV}$$

$$= \frac{0.007 \times 96\,487}{L} \text{ J}$$

Therefore, $\Delta E^0 = 0.007 \times 96\,487 = 675.4$ J. So for $T = 500$ K,

$$K_{P/P\dagger} = \left[\frac{3.022 \times 3.022}{2.015 \times 4.028} \right]^{\frac{3}{2}} \cdot \left[\frac{2 \times 2}{1 \times 1} \right] \cdot \left[\frac{85.4 \times 42.7}{64.0 \times 64.0} \right] \cdot \exp\left(-\frac{675.4}{8.314 \times 500} \right)$$

$$= 1.194 \times 4 \times 0.890 \times 0.850$$

$$= 3.614$$

Particular attention is drawn to the fact that the value of $K_{P/P^\dagger}$ (which is of course independent of the chosen value for $P^\dagger$) is almost completely accounted for by the differences in symmetry numbers of the molecules concerned. The type of reaction just studied is not the only one in which differences in symmetry numbers plays an outstanding part. In Section 7.9 we shall come upon a similar situation when we seek the reason for the preponderance of *meta*-xylene over *para* in an equilibrium mixture of the isomers.

The reaction $Cs(g) \rightarrow Cs^+(g) + \bar{e}$

This reaction is considered to illustrate the fact that a free electron can be treated as a monatomic gas molecule. Since the masses of the Cs atom and the Cs^+ ion are indistinguishable the translational partition functions for these species cancel and we are left with the equation

$$K_{P/P^\dagger} = \frac{(2\pi m_{(\bar{e})} kT)^{\frac{3}{2}} kT}{h^3 P^\dagger} \cdot \frac{g_{eo(\bar{e})} g_{eo(Cs^+)}}{g_{eo(Cs)}} \cdot \exp(-\Delta E^0/RT).$$

We have the following data:

  (a) the ionisation potential of caesium is 3.893 eV so that $\Delta E^0 = 3.893 \times 96\,487$ J,
  (b) the mass of the electron is 9.109×10^{-31} kg,
  (c) the free electron may have either of two possible spins so that $g_{eo(\bar{e})}$ is two. Also $g_{eo(Cs)}$ is two but $g_{eo(Cs^+)}$ unity.

Inserting the values of π, k and h and putting $P^\dagger = 1$ Nm^{-2} we have for $T = 3000$ K

$$K_{P/P^\dagger} = \frac{(2 \times 3.1416 \times 9.109 \times 10^{-31})^{\frac{3}{2}}}{(6.6256 \times 10^{-34})^3} \cdot (1.380 \times 10^{-23} \times 3000)^{\frac{5}{2}}$$

$$\times \exp\left[-\frac{3.893 \times 96\,487}{8.314 \times 3000}\right]$$

$$= 1.64 \times 10^7 \exp(-15.06)$$

$$= 4.72.$$

Attention is drawn to the fact that if we choose to assign the standard pressure the value $P^\dagger = 1$ atm $= 101.325$ kNm^{-2} the numerical value of $K_{P/P^\dagger}$ is reduced to 4.72/101 325 i.e. to 4.65×10^{-5}, or alternatively if we prefer to express the result in terms of K_P rather than $K_{P/P^\dagger}$ we can write

$$K_P = 4.72 \text{ Nm}^{-2} = 4.65 \times 10^{-5} \text{ atm.}$$

7.7   An Alternative Method for the Calculation of $K_{P/P\dagger}$

In the last three sections we showed how to calculate an equilibrium constant for a reaction between perfect gases directly from the molecular parameters $m, \sigma, \theta_r, \theta_v$ and g_{eo}, and in principle the method may be used however complicated be the reaction concerned. Inspection of equation (7.25) shows however that even for a relatively simple reaction the arithmetical exercise is often prolonged and tedious. Luckily another *completely equivalent* but simpler method is possible in many cases because of the wealth of statistical–mechanical data available in which most of the arithmetic has already been carried out.

In Section 7.3 it was shown that for any reaction

$$RT \ln K_{P/P\dagger} = -\Delta G^\dagger \tag{7.20}$$

where $-\Delta G^\dagger$ is the decrease in Gibbs function for a system resultant upon the conversion of a moles of A at its standard pressure $P^\dagger$ and b moles of B at the same pressure into c moles of C and d moles of D each at the same standard pressure $P^\dagger$. Thus for the reaction

$$aA + bB \cdots \rightleftharpoons cC + dD \cdots$$

$$RT \ln K_{P/P\dagger} = aG_A(T, P^\dagger) + bG_B(T, P^\dagger) \cdots - cG_C(T, P^\dagger) - dG_D(T, P^\dagger) \cdots \tag{7.29}$$

where $G(T, P^\dagger)$ is the Gibbs function *per mole* of each species at temperature T and at pressure $P^\dagger$.

In Section 5.9 we showed how to calculate the value of the molar quantity $-(G_o^T - E^0)/T$ for any perfect gas *at the standard pressure of one atmosphere*, and gave examples of the results of such calculations in Table 5.8, and mentioned the similar but not identical functions $-(G_o^T - H^{298.15})/T$ listed in *J.A.N.A.F. Tables of Thermochemical Data* published by the Dow Chemical Company. $K_{P/P\dagger}$ can be evaluated (through equation 7.29) using either set of data as will now be shown.

Thus

$$R \ln K_{P/P\dagger} = a\left(\frac{G_o^T - E^0}{T}\right)_{(A)} + b\left(\frac{G_o^T - E^0}{T}\right)_{(B)} \cdots$$
$$- c\left(\frac{G_o^T - E^0}{T}\right)_{(C)} - d\left(\frac{G_o^T - E^0}{T}\right)_{(D)} \cdots - \frac{\Delta E^0}{T} \tag{7.30}$$

where

$$\Delta E^0 = cE_{(C)}^0 + dE_{(D)}^0 \cdots - aE_{(A)}^0 - bE_{(B)}^0 \cdots \tag{7.31}$$

the numerical value for ΔE^0 being determined as shown in Section 7.4. This method is illustrated by the evaluation of $K_{P/P}{}^{\dagger}$ for the reaction

$$CO + H_2O \rightarrow CO_2 + H_2 \quad \text{at 600 K.}$$

Using the data of Table 5.8

$$R \ln K_{P/P\dagger} = \left(\frac{G_o^T - E^0}{T}\right)_{(CO)} + \left(\frac{G_o^T - E^0}{T}\right)_{(H_2O)} - \left(\frac{G_o^T - E^0}{T}\right)_{(CO_2)}$$

$$- \left(\frac{G_o^T - E^0}{T}\right)_{(H_2)} - \frac{\Delta E^0}{T}$$

$$= -189.21 - 178.94 + 206.02 + 122.09 + \frac{40\,422}{600}$$

the value $\Delta E^0 = -40\,422$ J having been calculated earlier in Section 7.4. Hence

$$R \ln K_{P/P\dagger} = 27.43 \text{ JK}^{-1} \quad \text{and} \quad K_{P/P\dagger} = 27.1.$$

Alternatively we may use the J.A.N.A.F. functions $-[(G_o^T - H^{298.15})/T]$. It follows then from equation (7.29) that

$$R \ln K_{P/P\dagger} = a\left(\frac{G_o^T - H^{298.15}}{T}\right)_{(A)} + b\left(\frac{G_o^T - H^{298.15}}{T}\right)_{(B)} \cdots$$

$$- c\left(\frac{G_o^T - H^{298.15}}{T}\right)_{(C)} - d\left(\frac{G_o^T - H^{298.15}}{T}\right)_{(D)} \cdots$$

$$- \frac{\Delta H^{298.15}}{T} \tag{7.32}$$

where $\Delta H^{298.15}$ is the heat of reaction at 298.15 K and is therefore obtainable from the molar heats of formation of the individual species by means of the equation

$$\Delta H^{298.15} = c\,\Delta H_f^{298.15}(C) + d\,\Delta H_f^{298.15}(D) \cdots - a\,\Delta H_f^{298.15}(A)$$
$$- b\,\Delta H_f^{298.15}(B) \cdots . \tag{7.33}$$

This method is illustrated by the evaluation of $K_{P/P\dagger}$ for the reaction

$$F_2 \rightleftharpoons F + F \quad \text{at 900 K.}$$

J.A.N.A.F. tables provide the following data for $T = 900$ K

$-\dfrac{G_o^T - H^{298.15}}{T}\Big/ \text{JK}^{-1}\,\text{mol}^{-1}$	F_2	F
	217.350	168.352
$\Delta H_f^{298.15}/\text{J mol}^{-1}$	0	78\,910

Hence

$$R \ln K_{P/P^\dagger} = \left(\frac{G_o^T - H^{298.15}}{T}\right)_{(F_2)} - 2\left(\frac{G_o^T - H^{298.15}}{T}\right)_{(F)} - \frac{\Delta H^{298.15}}{T}$$

$$= -217.350 + 2 \times 168.352 - \frac{157\,820}{900}$$

$$= -56.00 \, JK^{-1}$$

so that

$$K_{P/P^\dagger} = 1.188 \times 10^{-3}.$$

Recalling that the data used in this section is based on the standard pressure of one atmosphere we can if we wish express this result as

$$K_P = 1.188 \times 10^{-3} \, atm.$$

Attention is drawn to Exercise 7.6 in which $K_{P/P^\dagger}$ for this reaction is obtained from the values of the molecular parameters of F_2 and F using the method explained in Section 7.4.

7.8   Internal Rotation

In Section 5.2 we considered in detail the rotation of a rigid molecule about its centre of mass and mentioned that in the case of some polyatomic molecules relaxation of the condition of rigidity leads to an additional activity, that of the rotation of one part of the molecule with respect to another.

Let us consider in turn three molecules, ethylene, but-2-yne and ethane.

Of the $3n-6$ "vibrational" modes available to each molecule one is a *torsional mode* in which the two parts of the molecule are twisted relative to each other about the bond shown. In the case of ethylene the resistance of the double bond to torsional motion is high and the mode is a typical simple harmonic vibration which may be described by the equations of Section 5.3. In the case of but-2-yne however the resistance to torsional motion is so low that the two parts of the molecule are apparently capable of free internal rotation. The case of ethane is intermediate between the two extremes, and the molecule is said to exhibit *restricted internal rotation*. Internal rotations, in the cases in which they occur, increase the number of distinguishable configurations accessible to the molecule and so contribute to the thermodynamic properties of any assembly of which they are a part.

The contribution made by internal rotation to the properties of an assembly of gas molecules is small, but is sometimes very significant indeed. We have in mind the case of two isomers for each of which the sums of the contributions from translation, vibrational activity and rotation about the centres of mass are much the same, but for which the contributions from internal rotation differ. In such instances (as we shall see in the next section when we study such a case) the difference in internal rotational activity plays the dominant rôle in deciding the balance between one isomer and the other.

A complete statistical–mechanical description of internal rotation is out of place in an introductory text, and we shall attempt here only a superficial treatment without full theoretical justification. We shall consider first the case of free rotation in a molecule containing only *one* independently rotating group, *assuming that it is permissible to regard internal rotation and the rotation of the molecule as a whole about its centre of mass as independent activities*, secondly the circumstances in which this assumption is valid, thirdly the case in which the molecule contains more than one group capable of free rotation, and lastly the case in which the rotation is restricted.

(i) *Free internal rotation in the case of a single rotating group*

The quantum–mechanical energy levels associated with free internal rotation are given by the equation

$$\epsilon_{ir} = \frac{r^2 h^2}{8\pi^2 I_r} \tag{7.34}$$

where the quantum number r can assume the values $0, \pm 1, \pm 2, \ldots$, and where I_r, the *reduced* moment of inertia is defined by the equation

$$I_r = \frac{I_1 I_2}{I_1 + I_2} \tag{7.35}$$

I_1 and I_2 being the moments of inertia of each of the two parts about the axis of internal rotation. There is only one quantum state for each value of r so that the partition function for free internal rotation is given by the expression

$$f_{free} = \frac{1}{\sigma_{int}} \sum_{r=-\infty}^{r=\infty} \exp\left(-\epsilon_{ir}/kT\right). \tag{7.36}$$

The energy levels which are appreciably populated are small compared with kT even for temperatures close to absolute zero so that we can replace the summation term in (9.38) by the integral

$$\int_{-\infty}^{\infty} \exp\left(-\epsilon_{ir}/kT\right) dr,$$

from which we obtain the equation

$$f_{free} = \frac{1}{\sigma_{int}} \left(\frac{8\pi^3 I_r kT}{h^2}\right)^{\frac{1}{2}}. \tag{7.37}$$

The term σ_{int} figuring in the above is the appropriate symmetry number, being of course the number of otherwise distinct internal orientations which are indistinguishable due to the indistinguishability of like atoms. Obviously if a methyl group is the rotor, the value for σ_{int} is *three*, since we are unable to distinguish between the three equivalent orientations

$$-C\overset{\displaystyle H'}{\underset{\displaystyle H}{<}}H'', \quad -C\overset{\displaystyle H''}{\underset{\displaystyle H'}{<}}H \quad \text{and} \quad -C\overset{\displaystyle H}{\underset{\displaystyle H''}{<}}H'.$$

The contributions made by free internal rotation to the thermodynamic properties of an assembly of N molecules follow from (7.37).

$$E_{free}^T = NkT^2 d \ln f_{free}/dT = \tfrac{1}{2}NkT \tag{7.38}$$

$$C_{v\,free} = dE_{free}/dT = \tfrac{1}{2}Nk \tag{7.39}$$

and

$$S_{free}^T = Nk \ln f_{free} + NkTd \ln f_{free}/dT$$

$$= Nk \ln \left[\frac{1}{\sigma_{int}}\left(\frac{8\pi^3 I_r kT}{h^2}\right)^{\frac{1}{2}}\right] + \tfrac{1}{2}Nk \tag{7.40}$$

(ii)

We must now examine the circumstances in which it is permissible to regard free internal rotation and the rotation of the molecule as a whole about its centre of mass as independent activities and to describe the first by the equations given above. This is obviously permissible only if the energies associated with the rotation of the molecule about its centre of mass are unchanged when internal rotation occurs, and it is clear that this will be so *only if the principal moments of inertia of the molecule are unchanged in the course of internal rotation.* Strictly speaking this condition holds only in the case of molecules each part of which is a so-called *symmetric top.* Both groups in but-2-yne

$$
\begin{array}{cc}
\text{H} & \text{H} \\
\backslash & / \\
\text{H}-\text{C}-\text{C}\equiv\text{C}- \quad \text{and} \quad -\text{C}-\text{H} \\
/ & \backslash \\
\text{H} & \text{H}
\end{array}
$$

are symmetric tops whereas such groups as

$$
\begin{array}{cc}
\text{H} & \text{H} \\
\backslash & / \\
\text{H}-\text{C}-\text{C}\equiv\text{C}- \quad \text{and} \quad -\text{C}-\text{H} \\
/ & \backslash \\
\text{Cl} & \text{Cl}
\end{array}
$$

are not, and indeed it may be remarked that the agreement between the experimental values of C_v and of the calorimetric entropy for but-2-yne and the statistical–mechanical values obtained by taking into account contributions from translational and vibrational activities, the rotational activity of the molecule about its centre of mass and those resulting from internal rotation (as given by equations 7.39 and 7.40) leaves nothing to be desired.

Fortunately however the errors introduced by the use of the equations given above to cases in which both parts of the molecule are *not* symmetric tops are quite small *if the moment of inertia about the axis of internal rotation of one of the groups (i.e. either I_1 or I_2 in equation 7.35) is small compared with the principal moments of inertia of the molecule as a whole.* This means that these equations may be used in calculations carried out on many other molecules (such molecules as toluene H_3C–C_6H_5 and deutero-toluene H_2DC–C_6H_5) without introducing significant error.

(iii)

We now turn to the case in which the molecule contains more than one group capable of internal rotation. Obvious examples of such molecules are

$$
\textit{para-xylene} \quad
\begin{array}{cc}
\text{H} & \text{H} \\
\backslash & / \\
\text{H}-\text{C}-\bigcirc-\text{C}-\text{H} \\
/ & \backslash \\
\text{H} & \text{H}
\end{array}
$$

in which it may be supposed that both methyl groups can rotate relative to the benzene ring, and

CH$_3$

mesitylene

H$_3$C   CH$_3$

in which rotation of all three methyl groups may be supposed to be possible.

The situation here is that as long as the rotation of all groups remains free (we shall meet in the next section a case in which the rotation of one group in such a molecule prevents the free rotation of another) the total internal rotational partition function may be regarded as the product of the partition functions for the rotation of each group, and the total internal rotational contribution to the thermodynamic properties of the system the sum of the contributions made by each, *subject again to the condition that the moment of inertia of each group about the axis concerned is small compared with the principal moments of inertia as a whole.* Thus the total internal rotational contribution to the thermal capacity of *para*-xylene (supposing the rotation of each methyl group to be free) is Nk and the contribution to the entropy twice that given by equation (7.40), and the corresponding contribution to the thermal capacity of mesitylene $\frac{3}{2}Nk$ and the entropy contribution three times that given by (7.40).

(iv) *Restricted internal rotation*

As stated earlier, the agreement between the calorimetric value for the entropy of but-2-yne and the statistical–mechanical value obtained by taking into account the internal rotational contribution given by equation (7.40) is excellent. A similar comparison for ethane shows however that the statistical value (based on the assumption of *free* rotation) exceeds the calorimetric value by about $6\,\mathrm{J\,K^{-1}\,mol^{-1}}$. A comparison of the statistical–mechanical prediction for C_v with the experimental value shows a corresponding discrepancy. These results suggest that internal rotation in the case of ethane is somewhat restricted.

The forces which inhibit internal rotation in the ethane molecule may be supposed to arise from the repulsion between the C—H bonds in one methyl group and those in the other,[1] being greatest when the two sets of bonds are closest together as they will be when the two methyl groups are in the

---

[1] Presumably the corresponding forces in the case of but-2-yne are negligible because the two methyl groups are much further apart.

eclipsed conformation and least when the two groups are *staggered*. (Supposing that the molecule is standing on end, the staggered conformation is that in which all hydrogen atoms are visible and distributed symmetrically as shown in (a), and the eclipsed conformation that in which only the hydrogen atoms of one group are visible as shown in (b).)

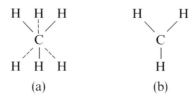

The potential energy for internal rotation in such a case is therefore a periodic function repeating itself every 120° ($2\pi/3$) because of the symmetry of the group. Such a function is shown in Figure 7.1 in which $V - V_{min}$ (where V_{min} is the potential energy corresponding to the staggered position) is plotted against the angle of rotation measured from that position, and in which $V - V_{min}$ is seen to reach a maximum value V_{max} when the groups are eclipsed. Obviously the closer the value of V_{max} to zero the more nearly free is the rotation and the greater the value of V_{max} the greater the restriction.

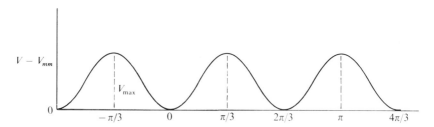

Figure 7.1   The potential energy between methyl groups in the ethane molecule.

Pitzer and his co-workers have prepared tables which permit the calculation of thermodynamic functions for various values of V_{max} for each possible value of f_{free}. These tables are given in Appendix 18 to Pitzer's monograph, *Quantum Mechanics*, (Prentice-Hall, 1953), and are based on a series of papers by Pitzer, Gwynn and Kilpatrick.

There is no theoretical basis for calculating the maximum potential V_{max}, but its value has been estimated in many cases by means of microwave and far-infra red spectroscopy. In such cases the Pitzer tables may be used directly to calculate the internal–rotational contribution to the thermodynamic properties of the compound. In cases in which the appropriate spectroscopic data are not available an estimate of V_{max} may still be obtained if accurate experimental values for the thermal capacity or entropy of the compound are known. In such a case the Pitzer tables are used to identify that value of V_{max} which most closely reconciles the calorimetric value and the statistical–mechanical value calculated assuming *free* internal rotation.

The foregoing gives only the briefest introduction to the mechanics of free and restricted internal rotation. For a much fuller account the reader is referred to Pitzer's monograph which gives the references to earlier papers on the topic, to a review by Dale, (*Tetrahedron* **22**, 3373 (1966)) and to a very recent review by Frankiss and Green (*Specialist Periodical Report, Chemical Thermodynamics*, Volume 1, The Chemical Society, 1973.)

7.9   The Thermodynamics of the Xylenes

As the last topic in this chapter we turn to a chemical equilibrium of a slightly different kind, that between the structural isomers *ortho*-, *meta*- and *para*-xylene. This topic is interesting because what is usually the most important contribution of all to the properties of a gas, the translational contribution, is the same for all structural isomers (because their molecular masses are the same), and in the case of the xylenes the vibrational contributions and the products of the three principal moments of inertia almost the same for each so that the differences in thermodynamic properties of the isomers are due almost completely to differences in what are usually rather minor matters, the symmetry factors and the differing degrees to which internal rotations in the molecules are restricted.

It is well known that a *mixture* of *ortho*-, *meta*- and *para*-xylene is produced during a Friedel–Crafts synthesis from toluene, and that any one of the xylenes will, in the presence of an aluminium halide and hydrogen halide rearrange to give some of each of the other two isomers, presumably by some sort of Wagner-Meerwein transformation. The composition of the equilibrium mixture (determined experimentally with, it may be remarked, very great difficulty) according to Norris and Vaala[1] is $16 \pm 10\%$ *ortho*, $65 \pm 10\%$ *meta* and $19 \pm 5\%$ *para*, and according to Pitzer and Scott[2] is $12 \pm 3\%$ *ortho*, $71 \pm 5\%$ *meta* and $17 \pm 2\%$ *para*. It is instructive first to examine the degree to which these figures are supported by the published values for the appropriate thermodynamic properties of the three species, and then to examine the statistical–mechanical origin of these values.

[1] J. F. Norris and G. T. Vaala, *J. Amer. Chem. Soc.* **61**, 1163 (1939).
[2] K. S. Pitzer and D. W. Scott, *ibid* **65**, 803 (1943).

The values for the standard molar heats of formation and standard molar entropies of the liquid xylenes at 25°C and those of the standard molar entropies of the gaseous xylenes at the same temperature shown in Table 7.1 are those reported in the *National Bureau of Standards Circular C461* of 1947 and are derived from three papers.[2,3,4]

Table 7.1

	$\Delta H_{f(liq)}/\text{kJ mol}^{-1}$	$S^\circ_{(liq)}/\text{JK}^{-1}\,\text{mol}^{-1}$	$S^\circ(g)/\text{JK}^{-1}\,\text{mol}^{-1}$
ortho	-24.44	246.48	352.75
meta	-25.42	252.17	357.69
para	-24.43	247.36	352.42

Classical thermodynamics shows that if A and B are components of a so-called *perfect* liquid mixture and are capable of interconversion the equilibrium constant for the reaction

$$A \rightleftharpoons B$$

is given by the equation

$$RT \ln K_x = RT \ln \frac{x_{B_{eq}}}{x_{A_{eq}}},$$

and since from (7.24), $RT \ln K_x = -\Delta G^\circ$,

$$RT \ln K_x = -\Delta H + T\Delta S^\circ$$
$$= -(\Delta H_{f_B} - \Delta H_{f_A}) + T(S^\circ_B - S^\circ_A) \qquad (7.41)$$

where $x_{A_{eq}}$ and $x_{B_{eq}}$ are the mole fractions of A and B at equilibrium, ΔH is the heat of reaction, ΔH_f the heat of formation of each species, S°_A the molar entropy of A as a pure liquid at temperature T and S°_B the corresponding quantity for B. The meaning of the word "perfect" in the phrase *perfect* liquid mixture is explained in Chapter 11.

A liquid mixture of the three xylenes approximates very closely to a perfect mixture, so that the values of $\Delta H_{f(liq)}$ and $S^\circ_{(liq)}$ given in Table 7.1 permit us to

[3] E. J. Prosen, W. H. Johnson and F. D. Rossini, *J. Research Natl. Bur. Standards* **36**, 455 (1946).

[4] W. J. Taylor, D. D. Wagman, M. G. Williams, K. S. Pitzer and F. D. Rossini, *ibid* **37**, 95 (1946).

estimate the composition of the equilibrium mixture of xylenes at 50°C by calculating the equilibrium constants for any two of the three equilibria

$$ortho \rightleftharpoons meta$$
$$\diagdown \qquad \diagup$$
$$para$$

Thus for the equilibrium

$$para \rightleftharpoons meta \quad \text{at } 25°C,$$

$$RT \ln K_x^{298.15} = -(\Delta H_{f(meta)} - \Delta H_{f(para)} + T(S^\circ_{meta} - S^\circ_{para})$$
$$= (990 + 1\,434) \text{ J},$$

so that

$$K_x^{298.15} = 2.66.$$

The classical thermodynamic relation[1] $d \ln K_x/dT = \Delta H/RT^2$ may now be used to estimate the value at 50°C and gives the value

$$K_x^{323.15} = 2.74.$$

Similarly for the equilibrium

$$ortho \rightleftharpoons meta$$

$$RT \ln K_x^{298.15} = -(\Delta H_{f\,(meta)} - \Delta H_{f\,(ortho)}) + T(S^\circ_{meta} - S^\circ_{ortho})$$
$$= (980 + 1696) \text{ J}$$

so that

$$K_x^{298.15} = 2.94 \quad \text{and} \quad K_x^{323.15} = 3.03.$$

For every 303 molecules of *meta* in the equilibrium mixture at 50°C we therefore have 100 molecules of *ortho* and 110 molecules of *para*, from which it follows that the equilibrium mixture contains 19.5% *ortho*, 59.1% *meta* and 21.4% *para*.

Before proceeding it must be remarked that less reliance can be placed on some of the values quoted in Table 7.1 than would be desired. Whilst the entropy values are probably of great accuracy (very good agreement having been obtained between the calorimetric values and those obtained as the result of spectroscopic measurements and the formulae of statistical mechanics) so that any error in the values of $(S^\circ_{meta} - S^\circ_{para})$ and $(S^\circ_{meta} - S^\circ_{ortho})$ is probably very small indeed, the same cannot be said of the values for ΔH_f. By whatever means a value is obtained for the heat of formation of

---

[1] The statistical–mechanical foundation of this equation is the subject of Exercise 7.8.

a compound of this kind, use has to be made at some stage of the experimental value for its heat of combustion. This for each of the xylenes is over 10^3 kJ mol^{-1} and the error in its determination can hardly be less than 0.5 kJ mol^{-1}; and this means that each of the values for ΔH_f given in Table 7.1 is uncertain to about 2%. For this reason Rossini and his co-workers (see reference 4 on page 166) consider that the composition of the equilibrium mixture at 50°C cannot be estimated better than the figures 20 ± 6% *ortho*, 58 ± 10% *meta* and 22 ± 8% *para*. The most that can be said therefore is that the value for the composition of the equilibrium mixture estimated using the principles of classical thermodynamics gives broad support to the experimental values obtained both by Norris and Vaala and by Pitzer and Scott. The lack of close agreement is however unimportant as far as we are concerned: the important point is the considerable preponderance of *meta* over *ortho* and *para*.

From the thermodynamic viewpoint the most interesting fact that emerges from the calculations shown above is that the quantities $(S^{\circ}_{meta} - S^{\circ}_{para})$ and $(S^{\circ}_{meta} - S^{\circ}_{ortho})$ are certainly as important, and probably rather more important than the quantities ΔH in determining the yield of each of the three isomers at equilibrium. It is therefore instructive to use our statistical–mechanical machinery to determine how these differences arise, in other words to inquire into the reasons why the molar entropy of *meta*-xylene is greater than that of *ortho* and *para*. We are of course able to identify the reasons for the differences in the entropies of the isomers only as perfect gases, but our findings will be directly applicable to the reactions in the liquid phase because the values of ΔS° and ΔH are almost the same in the liquid phase as they are for the gaseous reaction and the reasons for the differences the same.

The standard molar entropy of any one isomer is the sum of the following distinct contributions:

> S_t, the translational entropy which depends only on the molecular mass,
>
> S_e, the electronic entropy which we may assume to be zero,
>
> S_v, the vibrational entropy,
>
> S_r, the entropy contributed by the rotation of the molecule as a whole about its centre of mass

and

> $S_{i\cdot r}$, that contributed by the internal rotation of the methyl groups.

The various contributions to the standard molar entropies of the three gaseous xylenes at 25°C and 1 atm pressure are shown in Table 7.2. The translational contribution was calculated from equation (5.12),

$$S_t(298.15 \text{ K}, 1 \text{ atm}) = \tfrac{3}{2}R \ln M + 108.78 \text{ JK}^{-1} \text{ mol}^{-1}.$$

This contribution is of course the same for all three isomers. The vibrational contribution S_v is that calculated by Pitzer and Scott (see reference 2, page 165) from spectroscopic data. Because we wish to draw particular attention to the influence of the symmetry numbers of each species on the results obtained we report the entropy contributed by the rotation of the molecule about its centre of mass in two parts: $S_{r(i)}$ calculated from the equation

$$S_{r(i)} = R \ln f_r + RT \frac{d \ln f_r}{dT}$$

where

$$f_r = (\pi D)^{\frac{1}{2}} \frac{(8\pi^2 k T)^{\frac{3}{2}}}{h^3}$$

where D is the product of the three principal moments of inertia of the molecule, and $S_{r(\sigma)}$ from the equation

$$S_{r(\sigma)} = -R \ln \sigma,$$

where σ is the symmetry number for rotation of the molecule about its centre of mass. The appropriate value for *ortho*- and *meta*-xylene (each of which molecules possesses a single two-fold symmetry axis) is *two*, whilst that for *para*-xylene (which has three mutually perpendicular two-fold axes only two of which lead to additional orientations) is *four*, so that $S_{r(\sigma)}$ for *ortho* and *meta* is $-R \ln 2$ and for *para* $-R \ln 4$. The values S_{free} are the contributions which would result from the internal rotations of both methyl groups in each molecule were such rotations free, and are calculated from the equation

$$S_{\text{free}} = 2R \ln \left[\frac{1}{\sigma_{\text{int}}} \frac{(8\pi^3 I_r k T)^{\frac{1}{2}}}{h} \right] + R$$

being *twice* the value resulting from (7.40) because each molecule contains two rotating groups. The appropriate value for σ_{int} for each methyl group is *three* as explained in Section 7.8(i). S_{restr} is the amount by which S_{free} is reduced due to possible restrictions to free internal rotation. As explained in the previous section, the values for S_{restr} are obtained as the result of comparing the calorimetric value for the entropy with the statistical–mechanical value calculated assuming free internal rotation. Such calculations in the case of the xylenes were first carried out by Pitzer and Scott (see reference 2, page 165) and later revised slightly by Rossini and his co-workers (see reference 4). We see from the values of S_{restr} that internal rotation in the case of *meta*- and *para*-xylene is almost completely free, but that internal rotation of the methyl groups in *ortho*-xylene is much more seriously inhibited. This is of course just what we would expect by inspecting the structural formula for each isomer. The methyl groups in the case of *meta* and *para* are sufficiently far apart for the forces between the C–H bonds in one group and those in the other to be very small, but sufficiently close in the case of *ortho* to cause serious restriction. We now come to the point of these calculations.

Table 7.2 Contributions to the standard molar entropies of *ortho*-, *meta*- and *para*-xylene

	ortho	meta	para
$S_t/\mathrm{JK^{-1}\,mol^{-1}}$	166.90	166.90	166.90
$S_v/\mathrm{JK^{-1}\,mol^{-1}}$	52.17	50.09	51.29
$S_{r(i)}/\mathrm{JK^{-1}\,mol^{-1}}$	117.03	117.65	117.03
$S_{r(o)}/\mathrm{JK^{-1}\,mol^{-1}}$	−5.77	−5.77	−11.54
$S_{free}/\mathrm{JK^{-1}\,mol^{-1}}$	30.20	30.20	30.20
$S_{restr}/\mathrm{JK^{-1}\,mol^{-1}}$	−7.78	−1.38	−1.34
	352.75	357.69	352.42

The data in Table 7.2 put us in a position to identify the part played by each contribution to the value of ΔS° for each of the equilibria

$$para \rightleftharpoons meta \quad \text{and} \quad ortho \rightleftharpoons meta$$

in the gas phase. Thus for the equilibrium $para \rightleftharpoons meta$ we have to account for the value

$$\Delta S^\circ = S^\circ_{meta} - S^\circ_{para} = 357.69 - 352.42 = 5.27\ \mathrm{JK^{-1}}.$$

It follows from Table 7.2 that

$$\Delta S^\circ = \Delta S_t + \Delta S_v + \Delta S_{r(i)} + \Delta S_{r(\,)} + \Delta S_{free} + \Delta S_{restr}$$
$$= 0 - 1.20 + 0.62 + 5.77 + 0.12 - 0.04$$

so that the difference between the entropies of *meta*- and *para*-xylene is almost exactly accounted for by the greater degree of symmetry of the *para* molecule. Since we have already shown that the entropy term plays the larger part in the determination of the equilibrium point in the *meta* $\rightleftharpoons$ *para* equilibrium it is clear that the preponderance of *meta* over *para* is due mainly to the difference in symmetry of the two molecules concerned. This is an example of the general principle that *all other factors being the same the least symmetrical isomer is the one predominating in an equilibrium mixture.*

We now turn to the equilibrium *ortho* $\rightleftharpoons$ *meta*, for which we have to account for the value

$$\Delta S^\circ = S^\circ_{meta} - S^\circ_{ortho} = 357.69 - 352.75 = 4.94 \; JK^{-1}.$$

It follows from the data in Table 9.2 that

$$\Delta S^\circ = \Delta S_t + \Delta S_v + \Delta S_{r(i)} + \Delta S_{r(\;)} + \Delta S_{free} + \Delta S_{restr}$$
$$= 0 \quad - 2.08 + 0.62 \quad + 0 \quad + 0 \quad + 6.40$$

so that the differences between the molar entropies of *meta* and *ortho* is more than accounted for by the relatively large restriction to free internal rotation of the methyl groups in the *ortho* position. The preponderance of *meta* over *ortho* is therefore due almost entirely to this effect. This is an example of the general principle that *all other factors being the same the isomer in which internal restriction is least restricted is the one predominating.*

Before ending this chapter one further point should be made. In some cases a method may be used to determine the composition of an equilibrium mixture of structural isomers, which appears simpler than that used above, but which is in fact entirely equivalent to it.

Suppose that A, B and C are three isomers. We can then regard the collection of atoms constituting each molecule as being able to occupy

any one of a set of energy levels ϵ_A accessible only if the molecule has structure A,

or any one of a set of energy levels ϵ_B accessible only if the molecule has structure B,

or any one of a set of energy levels ϵ_C accessible only when the molecule has structure C.

It follows from the distribution law that at equilibrium the number of molecules choosing structure A will be given by the equation

$$\frac{N_A}{N_A + N_B + N_C} = \frac{f_A}{f_A + f_B + f_C} \tag{7.42}$$

and the numbers choosing structures B and C by similar equations, where f_A, f_B and f_C are the molecular partition functions for each of the three species. It follows that the ratios of the numbers of molecules of each species are given by the expression

$$N_A : N_B : N_C :: f_A : f_B : f_C \tag{7.43}$$

it being of course implicit in all expressions that the energy levels ϵ_A, ϵ_B and ϵ_C are measured from the same energy zero. This is achieved as we have seen many times before by expressing the electronic partition function as g_{e_0} and the vibrational partition function by an equation similar to (5.44') e.g.

$$f'_{v(A)} = \frac{\exp(D_{0(A)}/kT)}{\prod_{1,2,3,...}[1 - \exp(-\theta_{v(A)/kT})]}.$$

Since the translational partition function for all isomers is necessarily the same and the degeneracy of the electronic ground state presumably the same, expression (7.43) becomes

$$N_A:N_B:N_C::(f_r \cdot f'_v \cdot f_{i\cdot r})_A:(f_r \cdot f'_v \cdot f_{i\cdot r})_B:(f_r \cdot f'_v \cdot f_{i\cdot r})_C, \qquad (7.44)$$

and the ratios of the numbers on the left may be evaluated if the values for the molecular parameters $\sigma_r, \theta_r, D_0, \theta_v$ and I_r for each species are known, *and if all internal rotations occurring in each case are free*.

The reason why this rather more elegant method was not used to determine the composition of the equilibrium mixture of the xylenes is of course that the internal rotations of the methyl groups in these molecules are restricted so that explicit expressions for the terms $f_{i\cdot r}$ cannot be obtained.

EXERCISES

7.1 This exercise concerns the circumstances in which the equilibrium composition of a mixture may be calculated from the value of K_x. Suppose that A, B and C are perfect gases and that K_x for the reaction $A + B \rightleftharpoons C$ at a particular temperature and pressure is known. The mole fractions of each species at equilibrium are governed by two equations:

$$\frac{x_C}{x_A \cdot x_B} = K_x \qquad (i)$$

and

$$x_A + x_B + x_C = 1. \qquad (ii)$$

These equations are not sufficient to determine the value of each mole fraction; in other words (i) and (ii) are satisfied by an infinite number of sets of mole fractions depending on the initial amounts of A, B and C in the mixture and a particular set of values can be identified only if some additional information is given.

If for example we supply the information that initially the mixture contains n_A moles of A, n_B moles of B and n_C moles of C, and we suppose that in the course of the reaction to equilibrium n moles of A and (necessarily therefore) n moles of B react to form n moles of C, the equilibrium mole fractions are now given by the equations

$$x_A = \frac{n_A - n}{n_A + n_B + n_C - n}, \qquad x_B = \frac{n_B - n}{n_A + n_B + n_C - n},$$

$$x_C = \frac{n_C + n}{n_A + n_B + n_C - n}$$

and we have therefore the relationship

$$\frac{(n_C + n)(n_A + n_B + n_C - n)}{(n_A - n)(n_B - n)} = K_x$$

which contains only one 'unknown', the value of n. In these circumstances the composition of the equilibrium mixture can be determined.

Illustrate the above by carrying out the following exercise: Suppose that one mole of A and two moles of B are permitted to proceed to equilibrium at 5 atm pressure and that K_x for the reaction $A + B \rightleftharpoons C$ is 10.

Calculate

(a) the mole fractions of A, B and C at equilibrium,
(b) their partial pressures,
(c) the value of K_p.

A different type of reaction and one of great importance is that in which a species A dissociates into two species B and C, according, for example, to the equation $A \rightleftharpoons B + C$. We now have three relationships between the equilibrium mole fractions:

$$\frac{x_B \cdot x_C}{x_A} = K_x \tag{i}$$

$$x_A + x_B + x_C = 1 \tag{ii}$$

and

$$x_B = x_C. \tag{iv}$$

The value of K_x is now sufficient to determine the equilibrium mole fractions. Illustrate this by calculating

(d) the mole fractions of A, B and C at equilibrium,
(e) their equilibrium partial pressures,
(f) their value of K_p

for the case in which K_x for the reaction $A \rightleftharpoons B + C$ is 4 at a total pressure of 5 atm.

7.2 From the value of $K_{p/p\dagger}$ for the reaction $Cs \rightleftharpoons Cs^+ + \bar{e}$ at 3000 K given in Section 7.6 show that at that temperature and a total pressure of one atmosphere the degree of ionisation of caesium atoms is 6.9×10^{-3}.

7.3 (i) Calculate the value for $K_{p/p\dagger}$ for the reaction $H_2 \rightleftharpoons H + H$ at 3000 K using the following data:

$$m_{(H)} = 1.673 \times 10^{-27} \text{ kg}, \qquad g_{eo(H)} = 2, \qquad g_{eo(H_2)} = 1,$$

$$\sigma\theta_{r(H_2)} = 2 \times 85.4 \text{ K}, \qquad f_{v(H_2)} = 1,$$

ΔE^0 for the reaction is 432.2 kJ mol^{-1}.

(ii) From this result calculate the degree of dissociation of H_2 at 1 atm pressure and 3000 K.

7.4 Check the result obtained in Exercise 7.3(i) by calculating $K_{p/p\dagger}$ for the same reaction at 3000 K from the following data obtained from J.A.N.A.F. Tables:

	H_2	H
At $T = 3000$ K $-\dfrac{G_0^T - H^{298.15}}{T} \bigg/ \text{JK}^{-1} \text{ mol}^{-1}$	173.20	143.88
$\Delta H_f^{298.15}/\text{kJ mol}^{-1}$	—	218.0

7.5 (i) Calculate the value of $K_{p/p\dagger}$ for the reaction $H_2 \rightleftharpoons H + H$ at 1500 K using the data in Table 5.8.

(ii) If the value of $K_{p/p\dagger}$ is known at two temperatures, the value of ΔH for the reaction may be estimated from the van't Hoff equation

$$\frac{d \ln K_{p/p\dagger}}{dT} = \frac{\Delta H}{RT^2}.$$

(See exercise 7.8 for the statistical–mechanical verification of this equation.)

Use the value of $K_{p/p\dagger}$ for the reaction $H_2 \rightleftharpoons H + H$ at 3000 K obtained in Exercise 7.4 and the value for the reaction at 1500 K obtained in (i) above to obtain the mean value for ΔH for the reaction between 1500 and 3000 K. Compare this result with that obtained for the reaction at 298.15 K using the value for $\Delta H_{f(H)}$ at that temperature given in Exercise 7.4. Account for any difference.

7.6 In all the reactions studied so far the temperature has been such that we have rightly assumed that no error is introduced by supposing that all species are in their electronic ground state, i.e. by putting $f_e = g_{e_0}$. In cases where this assumption cannot be made we must replace the term g_{e_0} in such expressions as equations (6.41) by the expression

$$g_{e_0} + g_{e_1} \exp\left(- \Delta\epsilon_{e_1}/kT\right) + g_{e_2} \exp\left(- \Delta\epsilon_{e_2}/kT\right) + \cdots .$$

The present exercise requires such a substitution in the case of fluorine atoms.
You are given the following information:

(a) The molar mass of the F atom is 0.01899 kg. Its electronic ground state is four-fold degenerate, but this is accompanied by a two-fold degenerate state 404 cm^{-1} higher. No higher electronic states are accessible at the temperature of interest.

(b) The electronic ground state of the F_2 molecule is non-degenerate and the molecules are unexcited electronically at the temperature of interest.

(c) The internuclear distance in the F_2 molecule is 1.418×10^{-10} metre.

(d) The fundamental vibrational wave number for the molecule is 892 cm^{-1}.

(e) The heat of dissociation of F_2 at absolute zero is 154 kJ mol^{-1}.

Show

(i) that for the F atom

$$f_e = 4 + 2 \exp\left(- 581/T\right),$$

(ii) that for the F_2 molecule

$$I = 31.714 \times 10^{-47} \text{ kg m}^2.$$

Hence calculate θ_r from the formula $\theta_r = h^2/8\pi^2 Ik$.

(iii) that for the F_2 molecule $\theta_v = 1279$ K.

(iv) Hence calculate the value of $K_{p/p\dagger}$ for the reaction $F_2 \rightleftharpoons F + F$ at 900 K and check your result by comparing it with that value given in Section 7.7.

7.7 Textbooks on Organic Chemistry state that acetylene is readily converted into benzene by passing it through a red-hot tube in the absence of air. Show that this claim is probably justified by calculating the value of $K_{p/p\dagger}$ for the reaction $3C_2H_2 \rightleftharpoons C_6H_6$ at 1000 K using the data of Table 5.8.

7.8   One of the most useful relationships of classical thermodynamics is the van't Hoff equation quoted in Exercise 7.5(ii).
Derive this equation

(a) for the reaction

$$Cs_{(g)} \rightleftharpoons Cs^+_{(g)} + \bar{e}$$

using the equation for $K_{P/P\dagger}$ given in Section 7.6,

and   (b) for the more complicated reaction

$$H_2 + \tfrac{1}{2}O_2 \rightleftharpoons H_2O$$

using the expression for $K_{P/P\dagger}$ given in Section 7.4.

PART II

The Statistical Mechanics of the Canonical and Grand Canonical Ensembles

CHAPTER 8

The Canonical Ensemble and the System Partition Function

8.1 An Introduction to the Statistical Mechanics of Ensembles

The statistical mechanics described in previous chapters is impressive, but of limited application because its equations are applicable only to assemblies of *independent* particles. This limitation stems of course from the fact that the method used depends on our being able to express the energy of the system in terms of the 'private' energies of the constituent particles

$$E = n_0 \epsilon_0 + n_1 \epsilon_1 + \cdots + n_i \epsilon_i + \cdots$$

and it is not possible to frame such an equation for a system of particles between which inter-particle forces operate, and from the fact that although the equation

$$S = k \ln W$$

holds for *any* isolated system (and we shall show shortly that it holds for all practical purposes for any system whether isolated or not) it is possible to obtain expressions for W in terms of molecular parameters only for systems of independent particles.

In addition to this limitation the mechanics suffer from the disadvantage that it is necessary, every time a new problem is encountered, to set up an expression for the W value and to obtain the population numbers corresponding to its maximum term by means of Lagrange's method of undetermined multipliers. This procedure is lengthy, tedious, and (because the relations so far established apply only to isolated systems, systems which are closed and supposedly maintained at constant energy and at constant volume) particularly involved in those cases in which it is necessary to take into account the change in temperature which results from some processes in isolated systems, (we have in mind the passage from equation (7.5) to (7.8) in Section 7.1) or in those which entail the transfer of material from one part of the system to another.

Very much more powerful mechanics are available which go some way at least in removing the limitation and disadvantages mentioned above, and which make possible a statistical–mechanical attack on problems quite beyond the resources of the mechanics so far described. These are known as the *statistical mechanics of ensembles*.

What is meant by an *ensemble* is simply a very large collection of identical macroscopic systems which are allowed to interact. Thus whereas in previous chapters we considered *assemblies of particles*, we now consider *ensembles of systems*. We emphasise first, that no restriction is placed on the nature of the content of each system, in other words each system may be multiphase and contain dependent (interacting) particles, and secondly, that the number of systems in an ensemble is supposedly very large indeed, so large in fact that however minutely we find it necessary to sub-divide the ensemble we may safely assume that each sub-division consists of so many systems that we can apply the Stirling approximation formula to that number without sensible error.

The starting point of the mechanics of all ensembles is the same and entails the calculation of the properties of the ensemble and hence the average (mean) values for the properties of any one system in the ensemble.

The different ensembles are distinguished by the conditions governing the interaction of one system with another. In the simplest case each system is assumed to be *closed* and of fixed volume, but separated from its neighbours by diathermic walls so that all systems are in thermal equilibrium. This is called the *canonical* ensemble, and was first introduced by Gibbs. The most important quantity which emerges from its analysis is the so-called system partition function for constant V, T and N (here N denotes the amounts of all components) given by the expression

$$Q = \sum_i W(E_i) \exp(-E_i/kT), \qquad (8.1)$$

the summation extending over all possible values of E, and where $W(E_i)$ denotes the number of quantum–mechanical states characterised by energy E_i (for given N and V) and has precisely the same significance as that quantity figuring in earlier chapters.

It will be recalled that the most important property of the particle partition function f used in previous chapters is that the number of particles in an assembly in any of the states characterised by energy ϵ_i is given by the equation

$$\frac{n_i}{N} = \frac{g_i \exp(-\epsilon_i(kT)}{f}.$$

This property is shared by the system partition function Q (and by the other partition functions mentioned hereafter), $\mathcal{N}_i$ the number of systems in the canonical ensemble possessing energy E_i being given by the equation

$$\frac{\mathcal{N}_i}{\mathcal{N}} = \frac{W(E_i)\exp(-E_i/kT)}{Q}. \tag{8.2}$$

Expressions involving the system partition function are in general more useful than those involving the particle partition function hitherto used, because as emphasised above we now place no restriction on the nature of the systems in the ensemble. We observe however that since the systems are *closed*, the expressions of the canonical ensemble are likely to be no more *directly* applicable to a study of *open* phases within the system than are the equations of the isolated system of independent particles hitherto used.

This restriction is relieved to some extent by the so-called *grand canonical* ensemble also introduced originally by Gibbs. Here each system is considered to be of constant volume but *open*, and separated from its neighbours by diathermic permeable membranes, so that both material and energy can pass from one system to another. The most important quantity which emerges from the analysis of this ensemble is Z_G, the grand partition function for constant V, T and μ (μ representing the chemical potential of each component). For a system containing components A, B, ..., the grand partition function is given by the expression

$$Z_G = \sum_{N_A, N_B, \ldots E} W(N_A, N_B, \ldots, E_i) \exp[(N_A\mu_A + N_B\mu_B \cdots - E_i)/kT] \tag{8.3}$$

the multiple summation extending over all values of $N_A, N_B, \ldots$ and E, and where $W(N_A, N_B, \ldots, E_i)$ denotes the number of states characterised by energy E_i and content $N_A, N_B, \ldots$. Here again the partition function shares the property mentioned above, the number of systems in the ensemble possessing N_A molecules of A, N_B molecules of B, ... and in any of the states characterised by E_i being given by the equation

$$\frac{\mathcal{N}_{N_A, N_B, E_i}}{\mathcal{N}} = \frac{W(N_A, N_B, \ldots E_i)\exp(N_A\mu_A + N_B\mu_B \cdots - E_i)/kT}{Z_G} \tag{8.4}$$

The relations of the canonical ensemble are derived in later sections of this chapter, and those of the grand canonical ensemble are derived in Chapter 12.

There are several other ensembles.[1] We shall mention here only two, but shall neither construct their equations nor discuss their use. The first is an ensemble of *closed* systems in thermal and mechanical equilibrium with each other, so that any one system may change its volume at the expense of that of its neighbours. Its partition function for constant P, T and N is given by the expression

$$\Delta = \sum_{V,E} W(V_s, E_i) \exp\left[-(E_i + PV_s)/kT\right] \tag{8.5}$$

the double summation extending over all values of V and E, and where $W(V_s, E_i)$ denotes the number of states characterised by volume V_s and energy E_i, the number of systems in the ensemble occupying volume V_s and possessing energy E_i being given by the equation

$$\frac{\mathcal{N}_{s,i}}{\mathcal{N}} = \frac{W(V_s, E_i) \exp\left[-(E_i + PV_s)/kT\right]}{\Delta}. \tag{8.6}$$

This ensemble was first introduced by Guggenheim,[2] and has been discussed briefly by Brown[3] and more fully by Hill.[4]

What is probably the most powerful ensemble of all is that in which the systems are *open* and in thermal and mechanical equilibrium with each other. Its partition function for constant P, T, μ_A, μ_B, ... is given by the expression

$$\Lambda = \sum_{V,E,N_A,N_B,\ldots} W(N_A, N_B \ldots V_s, E_i) \exp\left[(N_A\mu_A + N_B\mu_B \cdots - E_i - PV_s)/kT\right] \tag{8.7}$$

in which the multiple summation extends over all possible values of E, V, N_A, N_B, ... and where $W(N_A, N_B, \ldots, V_s, E_i)$ denotes the number of states characterised by energy E_i, volume V_s and content N_A, N_B, ... Here again the number of systems in the ensemble containing N_A molecules of A, N_B molecules of B, ... occupying volume V_s and characterised by energy E_i is given by the equation

$$\frac{\mathcal{N}_{N_A,N_B,\ldots,V_s,E_i}}{\mathcal{N}}$$

$$= \frac{W(N_A, N_B, \ldots V_s, E_i) \exp\left[(N_A\mu_A + N_B\mu_B \cdots - E_i - PV_s)/kT\right]}{\Lambda} \tag{8.8}$$

---

[1] These are listed in Hill, *Introduction to Statistical Thermodynamics*, (Addison–Wesley, 1960 pp. 30–31).

[2] Guggenheim, *J. Chem. Phys.* **7**, 103 (1939).

[3] Brown, *Mol. Phys.* **1**, 68 (1958).

[4] Hill, *Statistical Mechanics*, (McGraw-Hill, 1958, Chapters 2 and 3).

This ensemble was also introduced originally by Guggenheim,[2] and some aspects of its use have been discussed by Hill.[4] It appears that the full potentialities of this ensemble have not yet been explored.

It is opportune to remark that, as will be demonstrated in the next section, an isolated system of independent particles may be regarded as a special case of the canonical ensemble, (differing from it only in so far that the canonical ensemble is an isolated assembly of independent *macroscopic systems*, whereas the systems with which we have so far been concerned are isolated assemblies of independent *particles*,) and for this reason the mechanics described in previous chapters are often known as the mechanics of a *micro-canonical ensemble*. We have not previously used this expression but shall, for the sake of brevity, use it when, in later sections, we need to refer to the earlier methods.

We proceed now to a detailed analysis of the canonical ensemble and to a description of some of its uses.

8.2   The Canonical Ensemble

Let us return to the consideration of the isolated assembly of localised independent particles of like kind studied in Chapter 2. The essential features of the model we need to recover are:

(a) that all particles are identical but distinguishable,

(b) that each may pass energy to its neighbours and may exist in a number of quantum states characterised by possible energies $\epsilon_0, \epsilon_1, \ldots, \epsilon_i, \ldots,$

(c) that the number of particles concerned is very large (otherwise we could not have used such mathematical conveniences as the fact that $\ln W_{\text{total}}$ may be replaced by $\ln W_{\text{max}}$, or that terms such as $\ln n!$ may be replaced by the Stirling expression $n \ln n - n$,)

and (d) that the energy and volume of the assembly is fixed as is the number of particles contained therein.

Keeping these features in mind let us now consider, not an isolated assembly of N identical, distinguishable particles, but an isolated ensemble of $\mathcal{N}$ identical closed macroscopic systems each occupying the same (fixed) volume V, and suppose that each system is separated from its neighbours by a diathermic barrier permitting the flow of heat from one to the other so that all systems in the ensemble are in thermal equilibrium. Denote the volume of the ensemble by the symbol $\mathcal{V}$, rnd its energy by the symbol $\mathcal{E}$.

Consideration shows that our *ensemble of systems* is a complete analogue of the *assembly of particles* discussed in Chapter 2 and may be described by a set of equations similar to those obtained in that chapter. Just as there we supposed that energy may pass from one particle to another so that energy levels $\epsilon_0, \epsilon_1, \ldots, \epsilon_i, \ldots$ are accessible to each, so we suppose that energy may pass from one macroscopic system to another so that each may be associated with any one of the energies $E_0, E_1, \ldots, E_i, \ldots$. Furthermore we may suppose that at any one moment

$\mathcal{N}_0$ systems may be characterised by energy E_0,

$\mathcal{N}_1$ by energy E_1,

$\vdots$

$\mathcal{N}_i$ by energy E_i,

$\vdots$

and at another moment

$\mathcal{N}'_0$ systems are characterised by energy E_0,

$\mathcal{N}'_1$ by energy E_1,

$\vdots$

$\mathcal{N}'_i$ by energy E_i,

$\vdots$

and so on, each set of numbers being governed by the equations

$$\mathcal{N} = \sum_i \mathcal{N}_i = \sum_i \mathcal{N}'_i = \cdots \tag{8.9}$$

and

$$\mathcal{E} = \sum_i \mathcal{N}_i E_i = \sum_i \mathcal{N}'_i E_i = \cdots \tag{8.10}$$

These equations are of course the analogues of (2.1) and (2.2) but what is the analogue of the equation

$$W_{\text{total}} = N! \prod_i \frac{g_i^{n_i}}{n_i!} + N! \prod_i \frac{g_i^{n_i}}{n'_i!} + \cdots? \tag{2.4}$$

The term g_i figuring in this equation is the number of (particle) quantum states characterised by the same energy ϵ_i. The corresponding quantity in the case of the macroscopic system possessing energy E_i and occupying volume V is therefore simply the W value which would be associated with the system were it isolated when possessing this energy and occupying this volume. It follows that the number of quantum states accessible to the ensemble is given by the equation

$$\Omega_{\text{total}} = \mathcal{N}! \prod_i \frac{W_i^{\mathcal{N}_i}}{\mathcal{N}_i!} + \mathcal{N}! \prod_i \frac{W_i^{\mathcal{N}_i}}{\mathcal{N}'_i!} + \cdots \tag{8.11}$$

where the term W_i has precisely the same significance as it has in previous chapters.

We can now proceed as we did in Chapter 2, that is to say:

(i) replace (8.11) by the expression

$$\ln \Omega_{\text{total}} \simeq \ln \Omega_{\text{max}} = \ln \mathcal{N}! \prod_i \frac{W_i^{\mathcal{N}_i^*}}{\mathcal{N}_i^*!} \tag{8.12}$$

where the symbol $\mathcal{N}_i^*$ denotes the set of population numbers (of systems within the ensemble) corresponding to the most probable distribution,

(ii) suppose that the set of most probable population numbers $\mathcal{N}_i^*$ are those satisfying the conditions

$$\delta \ln \Omega = \sum_i \left(\frac{\partial \ln \Omega}{\partial \mathcal{N}_i}\right) \delta \mathcal{N}_i = 0, \tag{8.13}$$

$$\sum_i \delta \mathcal{N}_i^* = 0, \tag{8.14}$$

and

$$\sum_i E_i \delta \mathcal{N}_i^* = 0, \tag{8.15}$$

(iii) combine these conditions by multiplying (8.14) by a multiplier α and (8.15) by a multiplier $-\beta$ adding to (8.13) and solve, finding that the number of systems associated with energy E_i is given by the equation

$$\mathcal{N}_i^* = \frac{\mathcal{N} W(E_i) \exp(-\beta E_i)}{\sum_i W(E_i) \exp(-\beta E_i)}, \tag{8.16}$$

the summation being carried out over all possible energies $E_0, E_1, \ldots,$ $E_i, \ldots$. We express the W value as $W(E_i)$ to emphasise that the degeneracy of each energy level depends on the energy.

We now define a quantity Q which we call the *partition function of the system* by the equation

$$Q = \sum_i W(E_i) \exp(-\beta E_i), \tag{8.17}$$

and so write

$$\mathcal{N}_i^* = \frac{\mathcal{N} W(E_i) \exp(-\beta E_i)}{Q}, \tag{8.18}$$

observe that the *chance* that at any moment a system be found associated with an energy E_j is given by the expression

$$p(E_j) = \frac{\mathcal{N}_j^*}{\mathcal{N}} = \frac{W(E_j)\exp(-\beta E_j)}{Q}, \tag{8.19}$$

and after substituting (8.18) into (8.12) and using the relation

$$\mathcal{S}_{\text{ensemble}} = k \ln \Omega_{\text{total}} \simeq k \ln \Omega_{\text{max}},$$

find that the entropy of the ensemble is given by the expression

$$\mathcal{S}_{\text{ensemble}} = k\mathcal{N} \ln Q + k\beta\mathcal{E}. \tag{8.20}$$

We can now, by using the same method as that used in Chapter 2, identify β as $1/kT$, where T is the equilibrium temperature of the ensemble, express Q as $\sum_i W(E_i)\exp(-E_i/kT)$, and show that

$$\mathcal{E}_{\text{ensemble}} = k\mathcal{N}T^2\left(\frac{\partial \ln Q}{\partial T}\right)_{\gamma}, \tag{8.21}$$

$$\mathcal{S}_{\text{ensemble}} = k\mathcal{N} \ln Q + k\mathcal{N}T\left(\frac{\partial \ln Q}{\partial T}\right)_{\gamma} \tag{8.22}$$

and

$$\mathcal{A}_{\text{ensemble}} = -k\mathcal{N}T \ln Q. \tag{8.23}$$

We now digress for a moment from the main purpose of this section to draw attention to the fact that since in our method we have, in effect, described each system in the ensemble by the same quantity β, and have identified β as a measure of temperature, we have shown that all systems in thermal equilibrium are characterised by the same temperature, and so have provided a statistical–mechanical foundation for the zeroth law of thermodynamics. The foundation of the zeroth law provided by a similar argument in Section 6.2 was limited to systems of independent non-localised particles. The foundation here provided applies to all systems whatever their nature.

Our progress to equation (8.23) follows precisely the same path as that beaten out in Chapter 2. We now depart from that path and record expressions for the *mean* energy, entropy and Helmholtz function for the systems in the ensemble. We denote these ensemble averages by the symbols $\bar{E}$, $\bar{S}$ and $\bar{A}$. Since

$$\mathcal{N}\bar{E} = \mathcal{E}_{\text{ensemble}},$$
$$\mathcal{N}\bar{S} = \mathcal{S}_{\text{ensemble}},$$

and

$$\mathcal{N}\bar{A} = \mathcal{A}_{\text{ensemble}},$$

it follows from equations (8.21), (8.22) and (8.23) that

$$\bar{E} = kT^2\left(\frac{\partial \ln Q}{\partial T}\right)_V, \tag{8.24}$$

$$\bar{S} = k \ln Q + kT\left(\frac{\partial \ln Q}{\partial T}\right)_V \tag{8.25}$$

and

$$\bar{A} = -kT \ln Q. \tag{8.26}$$

Before proceeding we should perhaps point out that in (8.21) and (8.22) we denoted constancy of volume by the suffix $\mathscr{V}$ (denoting constancy of volume of the *ensemble*) whereas in (8.24) and (8.25) we stipulate constancy of volume of the system. This is perfectly legitimate since we have previously stipulated constancy of both $\mathscr{V}$ and V.

Let us now focus attention on one particular system and consider its relation to the other systems in the ensemble. The only interaction between one system and the remainder is the flow of heat between them so that when the ensemble is in equilibrium, it and they are at the same temperature, and therefore the behaviour of our single system of interest is exactly the same whether it is part of an ensemble of identical systems in thermal equilibrium with it, or whether all the other systems are replaced by a thermostat at the same temperature. We may therefore dispense with all systems except the one on which we focus attention, and regard our system of interest as any closed constant volume system in thermal equilibrium with a thermostat at temperature T, and postulate:

(i) that its energy may assume the values $E_0, E_1, \ldots, E_i, \ldots$,

(ii) that the chance that at any moment it achieves an energy E_i is given by the expression

$$p(E_i) = \frac{W(E_i) \exp\left(-E_i/kT\right)}{Q}, \tag{8.27}$$

(iii) that over a period of time its mean energy is given by the equation

$$\bar{E} = kT^2\left(\frac{\partial \ln Q}{\partial T}\right)_V, \tag{8.28}$$

the mean value for its entropy by the equation

$$\bar{S} = k \ln Q + kT\left(\frac{\partial \ln Q}{\partial T}\right)_V \tag{8.29}$$

and the mean value for its Helmholtz function by the equation

$$\bar{A} = -kT \ln Q, \qquad (8.30)$$

the quantity Q being given by the equation

$$Q = \sum_i W(E_i) \exp(-E_i/kT). \qquad (8.31)$$

Attention is drawn to the fact that we have chosen to express Q as the sum over *accessible energy levels*. We could of course obtain the same quantity by summing over all *accessible quantum–mechanical states* in which case we should write

$$Q = \sum_{\text{states}} \exp(-E_i/kT) \qquad (8.32)$$

there being then $W(E_i)$ terms on the right for each value of E. As stated in the previous section, Q is called the system partition function for constant V, T and N.

8.3   Fluctuations

One aspect of the conclusions reached above calls for immediate comment: the postulate that the energy (or entropy or Helmholtz function) of a closed system of fixed composition and volume in equilibrium with a thermostat can assume more than one value. Indeed classical thermodynamics assumes that all properties of a system of fixed composition (here we mean a system not only of fixed chemical composition but one in which the amount of material in each phase is fixed as well) are determined if the values of any two are prescribed; in particular that its pressure and energy are determined completely by its temperature and volume. But if we seek the origin of this supposition we find that it apparently stems from such experimental evidence as that showing that the pressure of a given quantity of gas in a container of fixed volume is determined solely by its temperature, that the measured quantity of energy required to change a given mass of material from one state to another is always the same . . . and so on.

We have however already obtained some results which may make us wonder whether observations such as these are all that they seem. We have in mind the calculations made in Section 1.2 regarding the distribution of 10^{20} particles among labelled sites. We there found that although almost all complexions are encompassed by about 10^{10} configurations, so that there is considerable doubt as to the precise configuration in which the assembly would be found, no method of analysis is sufficiently accurate to distinguish any one of these possible configurations from the most probable configuration. We are therefore impelled to question whether, for example, the real instantaneous pressure of a gas *is* determined by its temperature and volume, and whether the apparent experimental observation that it is, is not in fact due to the *crudity* of the instruments by which it is measured. Whether the true situation is that the pressure is not in fact fixed but *fluctuates* from one value to another, and that we obtain the result which we do simply because we are unable to distinguish one actual measured value from another. In other words, whether the statement that the pressure of a gas is determined by its temperature and volume should not be replaced by the statement that if the pressure of a gas maintained at constant temperature and volume is measured, the chance that any result is obtained differing perceptibly from the most probable value is vanishingly small.

With this in mind, we see that in principle there is a considerable difference between a system maintained under such conditions that we *ensure* that its energy remains constant and one maintained under such conditions that we *ensure* constancy of temperature. In the latter case we do not ensure constancy of energy, and so must be prepared to accept provisionally the postulate that its energy may change and investigate the consequences of it. In short we accept the possibility of *fluctuations* in the energy of a closed system in a thermostat, and because of the obvious importance of this matter to the case in hand (in fact an understanding of the concept of fluctuations is prerequisite to the study of the mechanics of any of the various ensembles) we digress for a while in order to investigate this matter of fluctuations in some detail.

Let us suppose that a sample contains n like elements and that a set of energies $1, \ldots, i \ldots, r, \ldots$ are accessible to each. The set of population numbers $n_1, \ldots, n_i, \ldots, n_r, \ldots$ specifies the number of elements in each energy level and so represents the distribution of elements over energy levels, and since $n = \sum_i n_i$, the ratio $p_i = n_i/n$ represents the chance of finding a particular element with energy i.

If X denotes some physical property of the element depending on its energy, so that when the element has energy i the property acquires the value X_i, the mean value of the property X, denoted by $\overline{X}$, is given by the equation

$$\overline{X} = \frac{\sum_i n_i X_i}{n} = \sum_i p_i X_i, \tag{8.33}$$

and the deviation of any value X_i from the mean is given by $X_i - \overline{X}$. What we are really interested in is some measure of the distribution of values X_i about $\overline{X}$, because obviously the 'fine structure' of a sample in which all (or most) values of X_i are very close to the mean will be quite different from that of a sample in which all (or most) values depart appreciably from it. Whatever the distribution, the mean value of $X_i - \overline{X}$ is obviously zero, but the mean value of the quantity $(X_i - \overline{X})^2$, which is necessarily positive or zero, is just what is required, because it will be small for a distribution in which all (or most) values of X_i are very close to the mean and large for that in which all (or most) are very different from it.

We denote this quantity by the symbol $\overline{\delta X^2}$. It follows that

$$\overline{\delta X^2} = \overline{(X_i - \overline{X})^2} = \overline{X_i^2 - 2X_i\overline{X} + (\overline{X})^2}$$
$$= \overline{X_i^2} - (\overline{X})^2 \qquad (8.34)$$

since the mean value of $X_i\overline{X}$ is obviously $(\overline{X})^2$. This relation is quite general and applies whatever the nature of X. We are here particularly interested in the average deviation from the mean of the energy of a system in an ensemble of $\mathcal{N}$ such systems, the possible energies of each being $E_0, E_1, \ldots,$ $E_i, \ldots,$ so that

$$\overline{\delta E^2} = \overline{E^2} - (\overline{E})^2 \qquad (8.35)$$

where of course

$$\overline{E^2} = \frac{1}{\mathcal{N}}[E_0^2 + E_1^2 + \cdots + E_i^2 + \cdots] \qquad (8.36)$$

and

$$\overline{E} = \frac{1}{\mathcal{N}}[E_0 + E_1 + \cdots + E_i + \cdots]. \qquad (8.37)$$

It follows from (8.33) and (8.19) that

$$\overline{E} = \sum_i p_i E_i = \frac{\sum_i E_i W(E_i) \exp(-E_i/kT)}{Q}$$

and

$$\overline{E^2} = \sum_i p_i E_i^2 = \frac{\sum_i E_i^2 W(E_i) \exp(-E_i/kT)}{Q},$$

so that the expressions

$$Q\overline{E} = \sum_i E_i W(E_i) \exp(-E_i/kT) \qquad (8.38)$$

and

$$Q\overline{E^2} = \sum_i E_i^2 W(E_i) \exp(-E_i/kT) \qquad (8.39)$$

are identities. Differentiating the left-hand side of (8.38) with respect to T at constant volume, we see that

$$\left(\frac{\partial(Q\bar{E})}{\partial T}\right)_V = Q\left(\frac{\partial \bar{E}}{\partial T}\right)_V + \bar{E}\left(\frac{\partial \sum_i W(E_i) \exp(-E_i/kT)}{\partial T}\right)_V$$

$$= Q\left(\frac{\partial \bar{E}}{\partial T}\right)_V + \bar{E}\sum_i \frac{E_i}{kT^2} W(E_i) \exp(-E_i/kT)$$

$$= Q\left(\frac{\partial \bar{E}}{\partial T}\right)_V + \frac{Q(\bar{E})^2}{kT^2}. \tag{8.40}$$

Differentiating the right-hand side of (8.38) under the same conditions and making use of (8.39) we obtain the result

$$\frac{1}{kT^2}\sum_i E_i^2 W(E_i) \exp(-E_i/kT) = \frac{\overline{E^2}Q}{kT^2}. \tag{8.41}$$

Equating the right-hand sides of (8.40) and (8.41), dividing by Q and rearranging we see that

$$\overline{\delta E^2} = \overline{E^2} - (\bar{E})^2 = kT^2\left(\frac{\partial \bar{E}}{\partial T}\right)_V. \tag{8.42}$$

$(\partial \bar{E}/\partial T)_V$ is obviously the mean heat capacity at constant volume. Denoting this quantity by $\bar{C}_V$ we find that the average value of the root mean square deviation from the mean is given by the expression

$$(\overline{\delta E^2})^{\frac{1}{2}} = (kT^2\bar{C}_V)^{\frac{1}{2}} \tag{8.43}$$

and the relative value with respect to the mean by the expression

$$\frac{(\overline{\delta E^2})^{\frac{1}{2}}}{\bar{E}} = \frac{(kT^2\bar{C}_V)^{\frac{1}{2}}}{\bar{E}}. \tag{8.44}$$

This result is independent of the size or complexity of the system of interest. Let us first suppose that the system is a single atom (in which case the ensemble is simply an assembly of $\mathcal{N}$ independent atoms maintained at constant energy and volume and is identical with the assembly considered in Chapter 2). In this case C_V is of the order k (since molar heat capacities are of the order R i.e. Lk) and E is of the order kT (since molar energies are of the order RT i.e. LkT) and we see that $(\overline{\delta E^2})^{\frac{1}{2}}/\bar{E}$ is of the order unity. In other words the spread of systems over accessible energy levels is large, as indeed is illustrated by the distributions shown in Figures 3.3 and 3.4. Quite a different picture is obtained however when the system of interest is a macroscopic system containing N molecules where N is a very large number. In this case $\bar{C}_V$ is of the order Nk, and $\bar{E}$ of the order NkT and

$$\frac{(\overline{\delta E^2})^{\frac{1}{2}}}{\bar{E}} \simeq \frac{1}{N^{\frac{1}{2}}}, \tag{8.45}$$

so that for a macroscopic system 'just large enough to handle' where N is of the order 10^{20},

$$\frac{(\overline{\delta E^2})^{\frac{1}{2}}}{\bar{E}} \simeq 10^{-10}.$$

This result is illuminating indeed. It shows that although in principle any macroscopic system in an ensemble may assume any one of a vast number of energies (ranging from zero to the total energy of the ensemble) the chance that it acquires an energy distinguishable from the mean is vanishingly small.

We have already shown that the energies accessible to a system in an ensemble are the same as those accessible to it when immersed in a thermostat at the same temperature. The result we have just obtained means that the mean energy of a system in the ensemble given by equation (8.24) may be identified with the experimental (thermodynamic) value for the energy of the system immersed in a thermostat. The same reasoning applies to the entropy and Helmholtz function. In other words the experimental (thermodynamic) values for the energy, entropy and Helmholtz function for a macroscopic system in equilibrium with a heat bath at temperature T are given by the equations

$$E = kT^2 \left(\frac{\partial \ln Q}{\partial T} \right)_V, \tag{8.46}$$

$$S = k \ln Q + kT \left(\frac{\partial \ln Q}{\partial T} \right)_V \tag{8.47}$$

and

$$A = -kT \ln Q, \tag{8.48}$$

where

$$Q = \sum_i W(E_i) \exp(-E_i/kT). \tag{8.49}$$

Furthermore, the conclusion that the chance that a system in a thermostat achieves an energy E_j distinguishable from the thermodynamic value E is vanishingly small means that the value of $\ln Q$ obtained from equation (8.49) is indistinguishable from the logarithm of the maximum term in that sum, and we can write

$$\ln Q = \ln W(E) - \frac{E}{kT}. \tag{8.50}$$

The fact that expressions obtained using (8.49) are the same as those obtained using (8.50) is easily demonstrated. It will be shown later that for a system of N independent localised particles $\sum_i W(E_i) \exp(-E_i/kT)$ is identically equal to f^N where f is the particle partition function constructed in Chapter 2. It therefore follows from (8.47) that the entropy of such a system is given by the equations

$$S = k \ln Q + kT \left(\frac{\partial \ln Q}{\partial T} \right)_V$$

$$= Nk \ln f + NkT \left(\frac{\partial \ln f}{\partial T} \right)_V, \tag{8.51}$$

which result is, of course, the same as that obtained in Chapter 2.

If however we choose to use the value for Q given by (8.50),

$$\left(\frac{\partial \ln Q}{\partial T} \right)_V = \frac{E}{kT^2},$$

$$S = k \ln Q + kT \left(\frac{\partial \ln Q}{\partial T} \right)_V$$

$$= k \ln W(E) - \frac{E}{T} + \frac{E}{T}$$

$$= k \ln W(E),$$

which is the fundamental relationship from which all the equations of Chapter 2 were derived, and which leads to the same result as that given in (8.51).

It is worth dwelling for a moment on this result, because it means that the entropy of an isolated system calculated from the *micro-canonical* relations

$$S = k \ln W,$$

where

$$W = N! \prod_i \frac{g_i^{n_i}}{n_i!},$$

and where the population numbers n_i are given by the set of equations

$$n_i = \frac{Ng_i \exp(-\beta\epsilon_i)}{\sum_i g_i \exp(-\beta\epsilon_i)},$$

is indistinguishable from that of a system of the same composition maintained at temperature T, and given by the *canonical* relations

$$S = k \ln Q + kT \left(\frac{\partial \ln Q}{\partial T} \right)_V,$$

where

$$Q = \sum_j W(E_j) \exp(-E_j/kT),$$

so long as the parameter β figuring in the micro-canonical relations and the parameter T figuring in the relations of the canonical ensemble are connected by the equation

$$\beta = \frac{1}{kT}.$$

Thus the statistical–mechanical concept of entropy as a measure of the number of quantum–mechanical states accessible to the system, which arose from study of an isolated system, is equally applicable to a system in equilibrium with a thermostat.

Our study of fluctuations is complete for the time being, (we shall return to the topic when studying the grand canonical ensemble), but it is fitting to conclude this section with three remarks. The first is that it is easily shown (see Exercise 8.1) that the fluctuations in pressure of a perfect gas containing N molecules maintained at constant temperature and volume are governed by the expression

$$\frac{(\overline{\delta P^2})^{\frac{1}{2}}}{\overline{P}} \simeq \frac{1}{N^{\frac{1}{2}}},$$

so justifying our earlier surmise that if the pressure of a gas maintained at constant temperature and volume is measured the chance that any result is obtained which differs from the most probable value is vanishingly small. The second is that if a macroscopic system did *not* choose equilibrium values for its energy and pressure indistinguishable from the mean of all possible values, the development of physical science would have been quite different from what it was. Thus the gas laws and the first law of thermodynamics (to choose but two topics) would simply not *be*. The third point is to remind the reader of the results of some of the calculations carried out in Section 1.2. We there found, when considering the possible distributions of particles of two kinds, that the fraction of the total number of configurations encompassing almost all complexions is of the order $N^{-\frac{1}{2}}$, N being the number of particles concerned. It will now be realised that the fraction p/P figuring therein may be regarded as a measure of the root mean square deviation from the most probable distribution, and the fact that it proves experimentally to be of the order $N^{-\frac{1}{2}}$ is the result which would be expected from the theory of fluctuations.

8.4 The Canonical Ensemble and the Second Law of Thermodynamics

We now return to further consideration of equations (8.46) to (8.49). Of the first three equations, the most important is (8.48), because the Helmholtz function is the most useful thermodynamic property of a closed system maintained at constant temperature and volume, any variation in its state being governed by the classical thermodynamic equation

$$dA = -S\,dT - P\,dV + \sum_i \mu_i\,dN_i, \tag{8.52}$$

so that expressions for the entropy, pressure and chemical potentials of all components may be obtained by the appropriate differentiation. Thus

$$S = -\left(\frac{\partial A}{\partial T}\right)_{V,N} = k\ln Q + kT\left(\frac{\partial \ln Q}{\partial T}\right)_V$$

which is of course the same as (8.47),

$$P = -\left(\frac{\partial A}{\partial V}\right)_{T,N} = kT\left(\frac{\partial \ln Q}{\partial V}\right)_{T,N} \tag{8.53}$$

and

$$\mu_i = \left(\frac{\partial A}{\partial N_i}\right)_{T,V,N_j,\ldots} = -kT\left(\frac{\partial \ln Q}{\partial N_i}\right)_{T,V,N_j,\ldots}. \tag{8.54}$$

It is as well however, before proceeding, to demonstrate that the expression for A does indeed exhibit the most important characteristic of the classical thermodynamic Helmholtz function, the fact that for a closed system at constant temperature and volume the equilibrium state is that corresponding to the minimum value of A. This is easily done, because all that is necessary is to demonstrate that Q is a maximum in these circumstances.

By definition

$$Q = \cdots + W(E_h)\exp\left(-E_h/kT\right) + W(E_i)\exp\left(-E_i/kT\right)$$
$$+ W(E_j)\exp\left(-E_j/kT\right) + \cdots$$

there being one term for each possible energy level. Let us suppose that it were possible to enclose the system by adiabatic walls at the instant that its energy is E_h. $W(E_h)$ would then be the maximum number of complexions accessible to the system commensurate with its volume V and energy E_h. The same applies to each term. Thus each coefficient $\cdots W(E_h)$, $W(E_i)$, $W(E_j) \cdots$ is the maximum possible for each energy level, and since all possible energy levels are included in the expression for Q, Q *must be the maximum possible value for an equilibrium closed system occupying volume V at temperature T.* The quantity A defined by equation (8.48) *is* therefore the minimum value.

Since the equilibrium state of a closed system maintained at constant temperature and volume is that associated with the maximum value of Q, it follows, that in the language of the canonical ensemble, the feasibility of a process in those circumstances is the condition

$$\left(\frac{\partial Q}{\partial \xi}\right)_{T,V,N} > 0 \tag{8.55}$$

where ξ is the extent of reaction, or if we prefer

$$\left(\frac{\partial \ln Q}{\partial \xi}\right)_{T,V,N} > 0, \tag{8.56}$$

and the equilibrium state that determined by the condition

$$\left(\frac{\partial Q}{\partial \xi}\right)_{T,V,N} = 0, \tag{8.57}$$

or, if we prefer,

$$\left(\frac{\partial \ln Q}{\partial \xi}\right)_{T,V,N} = 0, \tag{8.58}$$

all of these expressions being the statistical–mechanical analogues of the classical condition

$$\left(\frac{\partial A}{\partial \xi}\right)_{T,V,N} \leqslant 0 \tag{8.59}$$

where the inequality sign refers to the feasibility of a process and the equality sign to the equilibrium state, and being therefore, the canonical ensemble expressions of the second law.

We are now in a position to make a preliminary comparison of the methods of the micro-canonical and canonical ensembles.

The fundamental quantity which arises in the statistical–mechanical theory of the isolated system is the number of quantum–mechanical states accessible to it, and the fundamental hypothesis of the theory is that any spontaneous process occurring in an isolated system is one leading to an increased spread over quantum–mechanical states, so that the conditions for spontaneous change and equilibrium are contained in the expression

$$\left(\frac{\partial \ln W}{\partial \xi}\right)_{E,V,N} \geqslant 0. \qquad \text{(i)}$$

The fundamental quantity which arises in the statistical–mechanical theory of the closed system maintained at constant temperature and constant volume is the partition function Q, and we have just demonstrated that any spontaneous process relating to such a system is one leading to an increase in Q, so that the conditions for spontaneous change and equilibrium are contained in the expression

$$\left(\frac{\partial \ln Q}{\partial \xi}\right)_{T,V,N} \geqslant 0. \qquad \text{(ii)}$$

The classical thermodynamic analogue of (i) is the expression

$$\left(\frac{\partial S}{\partial \xi}\right)_{E,V,N} \geqslant 0, \qquad \text{(iii)}$$

and that of (ii) the expression

$$-\left(\frac{\partial A}{\partial \xi}\right)_{T,V,N} \geqslant 0. \qquad \text{(iv)}$$

It is well known that in classical thermodynamics the identification of the equilibrium state by means of (iii) is often more difficult than the corresponding operation for a thermostatted system by means of (iv), because in applying (iii) we have to take into account the change in temperature resulting from most processes in isolated systems. The position in statistical mechanics is the same: the identification of the equilibrium state of an isolated system by means of (i) is often more difficult than the corresponding operation for a thermostatted system by means of (ii), the reason for this being the same as that given above. We shall find more than one illustration of this in later pages. The condition for chemical equilibrium was obtained in Section 7.1 using condition (i): in Exercise 8.3 we shall find that the condition can be established much more expeditiously using condition (ii). The conditions for phase equilibrium and for localised adsorption are established in Sections 9.1 and 10.2 by means of condition (ii): the more difficult task of their establishment by means of condition (i) is the subject of Exercises 9.2 and 10.3.

8.5   Properties of the Partition Function Q

It has been emphasised before that the relations of the canonical ensemble apply whatever the nature of the system concerned. They are therefore potentially more useful than the corresponding sets of relations developed in Chapters 2 and 4, which apply only to assemblies of independent particles.

It must however be recognised from the start that the power of the relations of the canonical ensemble is somewhat limited (in one direction at least) by the fact that we are able to *evaluate* the partition function Q only for systems of independent particules. (The same applies to the grand canonical partition function Z_G.) There is thus, for example, no question of our using equation (8.47) to calculate the molar entropy of liquid water, or indeed that of any system of interacting particles. The greater elegance and power of the relations of the canonical ensemble lie not in that direction at all, but depend on three features.

The first is that, as discussed in the last section, the use of these relations in investigating systems of independent particles is, in many cases, easier than that of those of the micro-canonical ensemble (essentially because the relations of the canonical ensemble are framed for a thermostatted rather than an isolated system). The second is that in certain circumstances it is possible to factorise Q into sub-system partition functions (Proposition I), and the third is that in other circumstances it is possible to factorise Q into partition functions for independent (or quasi-independent) degrees of freedom (Proposition II). It is on Proposition I that depend all applications of the mechanics of the canonical ensemble described in the remainder of this chapter and those described in Chapters 9 and 10, and on Proposition II that depends the statistical–mechanical study of liquid mixtures described in Chapter 11. These propositions are now established.

Proposition I

If the system is composed of two or more *distinguishable* closed sub-systems A, B, ... each of fixed volume, and independent in the sense that the energy of the system as a whole may be expressed as the sum of the energies of the parts (i.e. all potential energy terms relating the sub-systems are absent or can be ignored) so that

$$E = E_A + E_B + \cdots \qquad (8.60)$$

the partition function Q can be factorised into sub-system partition functions

$$Q = Q_A \cdot Q_B \cdots . \qquad (8.61)$$

Proposition I is established as follows: Suppose that sub-system A may assume any of a set of possible energies $E_{A_1}, E_{A_2}, \ldots, E_{A_i}, \ldots$, and that sub-system B may assume any of a set of energies $E_{B_1}, E_{B_2}, \ldots, E_{B_j}, \ldots$, the energy of the system as a whole being at any moment the sum of the energies of A and B. We may now construct partition functions for each sub-system,

$$Q_A = \sum_{\text{states}} \exp\left(-E_{A_i}/kT\right), \qquad (8.62)$$

and

$$Q_B = \sum_{\text{states}} \exp(-E_{B_j}/kT), \tag{8.63}$$

there being one term in each sum for each sub-system energy *state*. We have placed no restriction whatsoever on the size and complexity of the sub-systems so let us, for simplicity, consider the case in which we have two-sub-systems only, one with three energy states and one with two. Then

$$Q_A = \exp(-E_{A_1}/kT) + \exp(-E_{A_2}/kT) + \exp(-E_{A_3}/kT) \tag{8.64}$$

and

$$Q_B = \exp(-E_{B_1}/kT) + \exp(-E_{B_2}/kT), \tag{8.65}$$

and by multiplying together the right-hand sides of each equation we obtain the expression

$$\begin{aligned}
Q_A \cdot Q_B = {} & \exp[-(E_{A_1} + E_{B_1})/kT] + \exp[-(E_{A_1} + E_{B_2})/kT] \\
& + \exp[-(E_{A_2} + E_{B_1})/kT] + \exp[-(E_{A_2} + E_{B_2})/kT] \\
& + \exp[-(E_{A_3} + E_{B_1})/kT] + \exp[-(E_{A_3} + E_{B_2})/kT]
\end{aligned} \tag{8.66}$$

The possible energies for the system as a whole are those represented by the six sums shown in the exponents of (8.66). It follows that the expression for Q is precisely the same as that for the product $Q_A \cdot Q_B$ so that

$$Q = Q_A \cdot Q_B$$

which is Proposition I.

Proposition II

If one or more of the various degrees of freedom of the system may be assumed independent of the remainder (just as when considering an independent non-localised particle in Chapter 4 we supposed that its translational degrees of freedom are independent of its rotational, vibrational and electronic activities), so that the total energy of the system may be resolved into energies relating to each independent degree of freedom,

$$E = E_a + E_b + \cdots \tag{8.67}$$

the partition function Q can be factorised into partition functions relating to each independent degree of freedom, i.e.

$$Q = Q_a \cdot Q_b \cdot \cdots . \tag{8.68}$$

Suppose that the system with which we are concerned is an assembly of molecules each of which is capable of two independent activities a and b. Then each possible energy of the system, E_i, may be resolved into two components E_{ak} and E_{bj} so that

$$E_i = E_{ak} + E_{bj}, \qquad (8.69)$$

and $W(E_i)$, the total number of quantum–mechanical states associated with energy E_i may be factorised into two terms

$$W(E_i) = W(E_{ak}) \cdot W(E_{bj}) \qquad (8.70)$$

one being the degeneracy of energy level E_{ak}, and the other that of energy level E_{bj}. It follows that we may define partition functions Q_a and Q_b by the equations

$$Q_a = \sum_k W(E_{ak}) \exp\left(-E_{ak}/kT\right) \qquad (8.71)$$

and

$$Q_b = \sum_j W(E_{bj}) \exp\left(-E_{bj}/kT\right), \qquad (8.72)$$

so that the expression for Q, i.e.

$$Q = \sum_i W(E_i) \exp\left(-E_i/kT\right)$$

becomes

$$Q = \sum_k \sum_j W(E_{ak}) \cdot W(E_{bj}) \exp\left[-(E_{ak} + E_{bj})/kT\right]$$

$$= \sum_k W(E_{ak}) \exp\left(-E_{ak}/kT\right) \cdot \sum_j W(E_{bj}) \exp\left(-E_{bj}/kT\right)$$

$$= Q_a \cdot Q_b \qquad (8.73)$$

which is Proposition II.

It is upon this property of Q that many statistical–mechanical studies of systems of interacting particles depend, the point being that although when studying such systems we are unable to evaluate Q, we are, in some cases, able to evaluate some of its factors, and so able to offer partial solutions to problems quite beyond the resources of the mechanics described in earlier chapters. An example of this is provided in Chapter 11, where we study statistical–mechanical aspects of the chemistry of liquid mixtures.

8.6   The Condition for Mechanical Equilibrium

As the first example of the use of Proposition I, we establish the statistical–mechanical condition for mechanical equilibrium.

Consider a system occupying a fixed volume V, consisting of two sub-systems A and B, sub-system A containing a fixed quantity of gas, sub-system B containing a fixed but not necessarily the same quantity of gas, the two sub-systems being separated, as shown in Figure 8.1, by a frictionless piston. Suppose that the system is in thermal equilibrium with a heat bath at temperature T, and that the piston is first locked in some position so that the volume of sub-system A is V'_A and that of B, V'_B. We further suppose that initially the two sub-systems are *not* in mechanical equilibrium, so that when the piston is unlocked it will move in one direction or the other.

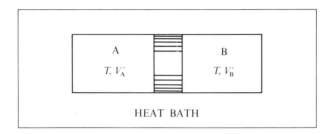

Figure 8.1

We denote the partition function of A in the initial state by Q'_A, noting that its value depends on T, V'_A and of course on its material content, and the partition function of B by Q'_B, noting that its value depends on T, V'_B and on its material content. The partition function for the complete system in its initial state is denoted by Q', the quantities Q', Q'_A and Q'_B being connected by the equation

$$Q' = Q'_A \cdot Q'_B.$$

We now suppose that the piston is unlocked and so free to seek that position corresponding to the state of mechanical equilibrium. It was shown in Section 8.4 that the equilibrium state of a closed system maintained at constant temperature is that at which its partition function is a maximum. The condition for mechanical equilibrium is therefore given by the equation

$$\left(\frac{\partial \ln Q}{\partial V_A}\right)_{T,V} = 0, \tag{8.74}$$

where V_A is the volume of sub-system A corresponding to the equilibrium position of the piston, or since

$$V = V_A + V_B$$

and

$$dV_A = -dV_B,$$

by the equivalent equation

$$\left(\frac{\partial \ln Q}{\partial V_B}\right)_{T,V} = 0. \tag{8.75}$$

Since the equilibrium value of Q may be expressed as the product of Q_A and Q_B, the conditions (8.74) and (8.75) are equivalent to the expression

$$\left(\frac{\partial \ln Q_A}{\partial V_A}\right)_T - \left(\frac{\partial \ln Q_B}{\partial V_B}\right)_T = 0,$$

or, since the temperature of each sub-system is the same, to the expression

$$kT\left(\frac{\partial \ln Q_A}{\partial V_A}\right)_T = kT\left(\frac{\partial \ln Q_B}{\partial V_B}\right)_T. \tag{8.76}$$

Inspection of (8.53) shows that the term on the left-hand side is the pressure of sub-system A, and that on the right, that of sub-system B.

8.7 The Evaluation of Q for a System of Independent Localised Particles of Like Kind

When establishing Proposition I no restriction was placed on the nature of the sub-systems except to suppose that they are distinguishable one from another and independent. We now suppose that the system consists of N identical, localised (and hence distinguishable) particles, the states accessible to each being characterised by the energies $\epsilon_0, \epsilon_1, \ldots, \epsilon_i, \ldots$. The system now considered is therefore identical with that studied in Chapter 2, except that there its energy was fixed whereas now we consider it maintained not at constant energy but at constant temperature.

It follows from (8.61) that

$$Q = Q_1 \cdot Q_2 \cdot \cdots \cdot Q_N \tag{8.77}$$

where Q_1 is the partition function of one particle (sub-system), Q_2 that of the second $\cdots \cdot Q_N$ that of the Nth. Each sub-system partition function is given by the expression $\sum_i \exp\left(-\epsilon_i/kT\right)$, there being one term in the sum for each possible energy state, i.e. g_i terms for each energy level ϵ_i, so that each sub-system partition function is identical with f, the particle partition function defined in Chapter 2. It follows from (8.77) that

$$Q = f^N. \tag{8.78}$$

It will be recalled that the validity of this result has already been tested in the derivation of equation (8.51).

8.8   The Evaluation of *Q* for a System of Independent Non-Localised Particles of Like Kind

In establishing Proposition I we postulated that the sub-systems are independent but *distinguishable*, and therefore a system of N identical, non-localised (*indistinguishable*) particles falls without the conditions under which (8.61) was deduced, and this relationship no longer applies. The reason for this is easily seen. For a system containing two *distinguishable* particles A and B, each capable of existing in two energy states,

$$Q_A = \exp\left(-\epsilon_{A_1}/kT\right) + \exp\left(-\epsilon_{A_2}/kT\right), \tag{8.79}$$

$$Q_B = \exp\left(-\epsilon_{B_1}/kT\right) + \exp\left(-\epsilon_{B_2}/kT\right) \tag{8.80}$$

and

$$Q = Q_A \cdot Q_B = \exp\left[-(\epsilon_{A_1} + \epsilon_{B_1})/kT\right] + \exp\left[-\epsilon_{A_2} + \epsilon_{B_2})/kT\right].$$
$$+ \exp\left[-(\epsilon_{A_2} + \epsilon_{B_1})/kT\right] + \exp\left[(-\epsilon_{A_2} + \epsilon_{B_2})/kT\right]. \tag{8.81}$$

But if the particles are *indistinguishable* no labels A and B are appropriate, and the state denoted above by $(\epsilon_{A_1} + \epsilon_{B_2})$ is indistinguishable from that denoted by $(\epsilon_{A_2} + \epsilon_{B_1})$ and in the expression for Q should appear only once. The expression

$$Q = \exp\left[-(\epsilon_1 + \epsilon_1)/kT\right] + \exp\left[-(\epsilon_1 + \epsilon_2)/kT\right]$$
$$+ \exp\left[-(\epsilon_2 + \epsilon_2)/kT\right] \tag{8.82}$$

would thus be correct for particles obeying Bose–Einstein statistics in which more than one particle is allowed in each quantum state. It however the particles obey Fermi–Dirac statistics even (8.82) is incorrect because the first and third terms are forbidden since they both correspond to two particles in the same state, and for such a system

$$Q = \exp\left[-(\epsilon_1 + \epsilon_2)/kT\right]. \tag{8.83}$$

We shall make no attempt to deduce the correct expression for Q, but accept that in those circumstances in which the degeneracy of each particle energy level greatly exceeds the number of particles therein (so that the distinction between Bose–Einstein and Fermi–Dirac statistics may be ignored, just as it was in Chapter 4), the system partition function is given by the expression

$$Q = \frac{f^N}{N!}. \tag{8.84}$$

This relation fails only at very low temperatures, (see the footnote on page 80).

It is left to the reader to show that substitution of $f^N/N!$ for Q in equations (8.46) to (8.48) leads to expressions for the properties of a perfect gas which are identical to those given in equations (4.32) to (4.34).

8.9 The Evaluation of Q for Homogeneous Systems of Independent Particles of More Than One Kind

The partition function for a mixture of perfect gases containing N molecules of A, M molecules of B, ... is given by the equation

$$Q = \frac{f_A^N \cdot f_B^M}{N! \; M!} \cdot \ldots \tag{8.85}$$

this expression being valid in the same circumstances as is (8.84). It is left to the reader to show that substitution of the right-hand side of (8.85) for Q in equations (8.46) to (8.48) leads to expressions for the properties of a perfect gas mixture identical to those given in Section 6.2.

The construction of an expression for Q for a mixture of localised particles of two kinds A and B, supposing that there are N particles of A and M of B, lies at the heart of the problem discussed in Chapter 11. Here we state without explanation that if the particles are truly independent *and of the same size and shape*, the partition function is given by the expression

$$Q = \frac{(N + M)!}{N! \, M!} \cdot f_A^N \cdot f_B^M, \tag{8.86}$$

the term $(N + M)!/N! \, M!$ representing the number of distinct geometric arrangements of N localised particles of one kind and M of another.

8.10 The Construction of the Partition Function for Closed Systems Containing Open Phases

The relations of the canonical ensemble are framed for *closed* systems maintained at constant volume and constant temperature, and are not therefore directly applicable to *open* phases within a system. The method by which the partition function is constructed for a closed system containing open phases is demonstrated by considering the simplest possible case, that of a system consisting of two phases, one an open sub-system of independent localised particles, the other of an open sub-system of independent non-localised particles of the same kind.

Suppose that a crystalline solid containing N identical, independent, localised particles is introduced into a closed vessel of fixed volume, and that M particles pass into the gas phase, the system being maintained at temperature T by means of a heat bath. For simplicity, make one assumption, which although not strictly true introduces no significant error: that is that the change in volume of the solid phase resulting from the sublimation process is negligible. This assumption implies that the volume accessible to the gas molecules is independent of the value of M, as therefore are the energy levels accessible to them and the value of the particle partition function $f_{(g)}$.

It is simple to frame a partition function for the system *for each value of* M, because by assigning a definite value to M we are, in effect, imposing closure on each phase. Thus, it follows from (8.61), (8.78) and (8.84) that for the value $M = 1$,

$$Q_1 = Q_{(g_1)} \cdot Q_{(s)(N-1)} = \frac{f_{(g)}}{1!} \cdot f_{(s)}^{(N-1)},$$

for the value $M = 2$,

$$Q_2 = Q_{(g)_2} \cdot Q_{(s)(N-2)} = \frac{f_{(g)}^2}{2!} \cdot f_{(s)}^{(N-2)},$$

for the value $M = n$,

$$Q_n = Q_{(g)_n} \cdot Q_{(s)(N-n)} = \frac{f_{(g)}^n}{n!} \cdot f_{(s)}^{(N-n)},$$

and so on.

It follows from equation (8.58) that M^*, the most probable (equilibrium) value of M is that leading to the maximum value for Q_M, and so is that determined by the equation

$$\left(\frac{\partial \ln Q_{M^*}}{\partial M^*}\right)_{T,V,N} = 0. \tag{8.87}$$

We have immediately,

$$\ln Q_{M^*} = M^* \ln f_{(g)} - M^* \ln M^* + M^* + (N - M^*) \ln f_{(s)}, \tag{8.88}$$

so that

$$\left(\frac{\partial \ln Q_{M^*}}{\partial M^*}\right)_{T,V,N} = \ln f_{(g)} - \ln M^* - \ln f_{(s)}. \tag{8.89}$$

It follows from (8.87) that the equilibrium value of M is that given by the equation

$$\frac{f_{(g)}}{M^*} = f_{(s)}. \tag{8.90}$$

The method established above is of considerable importance. We shall return to it in the next chapter, where we consider the general condition for phase equilibrium, and again in Chapter 10, where we consider a phase equilibrium of a special kind, that governing the adsorption of a gas on a crystalline surface.

EXERCISES

8.1 Consider a perfect gas containing N molecules maintained at temperature T and volume V. Making use of the fact that the pressure of a perfect gas is related to its energy by the equation

$$PV = \tfrac{2}{3}E,$$

use (8.45) to show that the relative value of the root mean square deviation of pressure with respect to the mean is given by the expression

$$\frac{(\overline{P^2})^{\frac{1}{2}}}{\overline{P}} \simeq N^{-\frac{1}{2}}. \tag{8.E1}$$

Although this equation is correct, are there some doubts that the *measured* value for the pressure of a highly diffuse gas might differ appreciably from the mean? The point to which we draw attention is the fact that although measurements relating to the energy of a system (measurements of heat capacities, heats of reaction and so on) are necessarily made on large samples of the material, a pressure gauge may consist of a probe of rather small area of contact. This means that the gauge is subject to collisions by relatively few molecules. Is the relative value of the deviation from the mean observed by the gauge given by equation (8.E1) or by the equation

$$\frac{(\overline{P^2})^{\frac{1}{2}}}{\overline{P}} \simeq n^{-\frac{1}{2}},$$

where n is the number of molecules striking the probe area during the time taken for a reading to be made?

8.2 It was shown in Section 8.10 that the number of gas molecules in equilibrium with an assembly of independent localised particles is given by the equation

$$\frac{f_{(g)}}{M^*} = f_{(s)}. \tag{8.90}$$

Hence, making use of the fact that the partition function for an independent, non-localised particle is proportional (through the translational partition function) to the volume accessible to it, so that

$$f_{(g)} = f'_{(g)} \cdot V, \tag{8.E2}$$

show that the equilibrium vapour pressure of a crystalline solid is given by the equation

$$P = kT \frac{f'_{(g)}}{f_{(s)}}. \tag{8.E3}$$

8.3 It was implied in the opening paragraphs of this chapter that the use of the relations of the canonical ensemble is less laborious than that of those of the mechanics described in earlier chapters. Illustrate this by returning to the problem discussed in Section 7.1, that of the establishment of the condition for chemical equilibrium in a perfect gas reaction.

Consider a system containing a mixture of perfect gases A, B, C, D, ... (maintained at constant V and T) and suppose that a reaction

$$aA + bB \cdots \rightleftharpoons cC + dD \cdots$$

is possible. Show that the value of Q for a mixture containing N_A molecules of A, N_B molecules of B, ... is given by the expression

$$Q_{N_A, N_B, N_C, N_D, \ldots} = \frac{f_A^{N_A}}{N_A!} \cdot \frac{f_B^{N_B}}{N_B!} \cdot \frac{f_C^{N_C}}{N_C!} \cdot \frac{f_D^{N_D}}{N_D!} \cdot \quad \cdots \qquad (8.E4)$$

and that the equilibrium values $N_A^*, N_B^*, N_C^*, N_D^*, \ldots$ are those satisfying the equation

$$\left(\frac{\partial \ln Q_{N_A^*, N_B^*, N_C^*, N_D^*, \ldots}}{\partial \xi} \right)_{T,V} = 0 \qquad (8.E5)$$

where (see equation 7.3)

$$d\xi = -\frac{dN_A^*}{a} = -\frac{dN_B^*}{b} = \frac{dN_C^*}{c} = \frac{dN_D^*}{d} = \cdots .$$

Hence show that the condition for chemical equilibrium is given by equations (7.9) and (7.10).

It will be found that the final equations are obtained much more expeditiously than they were in Chapter 7, because here we are concerned with a system maintained at constant temperature, whereas there, we were concerned with a system of constant energy, and had to take into account the change in temperature resulting from the reaction, and its effect on the values of $f_A, f_B, f_C, f_D, \ldots$.

8.4   In the monograph *Applications of Statistical Mechanics* (Clarendon Press, Oxford, 1965) Guggenheim remarks "... the problem 'given the energy of a system evaluate its temperature' is much less tractable than its converse 'given the temperature of a system evaluate its energy' ...".

Discuss the bearing of this quotation on the subject of this chapter, and in particular on the contents of Section 8.4.

CHAPTER 9

Phase Relations

We have already shown (see Sections 6.2 and 8.2) that all parts of a system in equilibrium are characterised by the same temperature, and, (see Section 8.6) that all parts are characterised by the same pressure. We now consider a system composed of open phases, and establish the condition for the equilibrium distribution of components between one phase and another.

9.1 The Condition for Phase Equilibrium

Consider first a system composed of two closed sub-systems A and B, and suppose that sub-system A occupies a fixed volume V_A and consists of an unspecified quantity of material but contains among its other components M'_j molecules of component j, and that sub-system B occupies a fixed volume $V - V_A$ and contains among its other components $N - M'_j$ molecules of j. Suppose that the two sub-systems reach thermal equilibrium in a heat bath at temperature T, but are separated from each other by a diathermic barrier *impermeable to all components*, as shown in Figure 9.1.

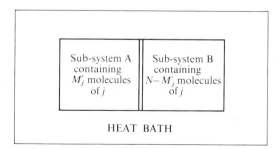

Figure 9.1

209

We can now allot partition functions Q'_A, Q'_B and Q' to each sub-system and to the system as a whole, these quantities being connected, in accordance with Proposition I of Chapter 8, by the equation

$$Q' = Q'_A \cdot Q'_B. \tag{9.1}$$

Now suppose that the barrier separating the two sub-systems is replaced by a membrane permeable only by component j. $N + 1$ configurational states are now possible to the system, these corresponding to all values of M_j from zero to N, since all or some of the M'_j molecules originally in sub-system A may pass into sub-system B, or all or some of the $N - M'_j$ molecules originally in sub-system B may pass into sub-system A.

To each of these $N + 1$ configurational states we may allot sub-system partition functions $Q''_A, Q''_B; Q'''_A, Q'''_B, \ldots$ corresponding to the number of molecules of j in each part, so that for each value of M_j we have the expression

$$Q_{M_j} = Q_{A(M_j)} \cdot Q_{B(N - M_j)}. \tag{9.2}$$

We see, in accordance with equation (8.58), that the most probable (equilibrium) distribution of j between the two sub-systems is that leading to the maximum value of Q_{M_j}, and so is that satisfying the equation

$$\left(\frac{\partial \ln Q_{A(M_j)} \cdot Q_{B(N - M_j)}}{\partial M_j} \right)_{T,V,N} = 0,$$

or, since

$$dM_j = -d(N - M_j),$$

and V_A is fixed, that satisfying the equation

$$\left(\frac{\partial \ln Q_{A(M_j)}}{\partial M_j} \right)_{T,V_A} = \left(\frac{\partial \ln Q_{B(N - M_j)}}{\partial(N - M_j)} \right)_{T,(V - V_A)} \tag{9.3}$$

or, since the temperature of each sub-system is the same, that satisfying the equation

$$-kT \left(\frac{\partial \ln Q_{A(M_j)}}{\partial M_j} \right)_{T,V_A} = -kT \left(\frac{\partial \ln Q_{B(N - M_j)}}{\partial(N - M_j)} \right)_{T,(V - V_A)}. \tag{9.4}$$

Inspection of equation (8.54) shows that the expression on the left-hand side of (9.4) is the chemical potential of j in sub-system A, and that that on the right is that of j in sub-system B. We conclude that the condition for equilibrium between two open sub-systems (phases) with respect to component j is that

$$\mu_{j(A)} = \mu_{j(B)}. \tag{9.5}$$

Analogous conditions can be obtained in respect of all other components, and for systems composed of more than two phases. We conclude that the chemical potential of any component present in more than one phase in an equilibrium system is the same in each. Those familiar with classical thermodynamics will know that this conclusion provides the thermodynamic foundation for much of physical chemistry.

The remainder of this chapter is spent in describing some of the applications of condition (9.5) to one-component systems. Some aspects of phase relations with respect of two-component systems are considered in Chapter 11.

9.2   Systems of Independent Particles

In this section and the next we consider systems of independent particles because it is only in such cases that we can express the chemical potentials in terms of molecular parameters.

Suppose therefore, that a crystalline solid containing N identical, independent, localised particles is introduced into a vessel of fixed volume and maintained at temperature T, and that when equilibrium is established M particles have passed into the gas phase, the volume accessible to them being V. We then have

$$\mu_{(s)} = \mu_{(g)}. \tag{9.6}$$

It follows from the equations of Chapters 2 and 3 that $\mu_{(s)}$ is given by the expression $-kT \ln f_{(s)}$, and from the equations of Chapters 4 and 5 that $\mu_{(g)}$ is given by the expression $-kT \ln f_{(g)}/M$, so that

$$-kT \ln f_{(s)} = -kT \ln \frac{f_{(g)}}{M}, \tag{9.7}$$

from which it follows that

$$f_{(s)} = \frac{f_{(g)}}{M}; \tag{9.8}$$

which is the same result as that obtained in Section 8.10. We observe that neither N, the total number of particles in the system, nor $N - M$, the number remaining in the solid phase, figure in equations (9.7) and (9.8).

Equation (9.8) provides a particularly simple picture of the progress of a sublimation process. The value of $f_{(s)}$ for a given molecule depends only on its molecular volume and on the temperature, and the value of $f_{(g)}$ on the temperature and the total volume accessible to the molecule. Both values of $f_{(s)}$ and $f_{(g)}$ are therefore determined by the conditions imposed on the system.

Sublimation proceeds until the number of molecules entering the gas phase reaches the value of the quotient $f_{(g)}/f_{(s)}$.

An expression for the equilibrium vapour pressure of the solid follows immediately from (9.8). Since the partition function of an independent, non-localised particle is proportional (through the translational partition function) to the volume accessible to it, so that

$$f_{(g)} = f'_{(g)} \cdot V,$$

where $f'_{(g)}$ is a function only of temperature, and since, from (5.75), the pressure of an assembly of M independent, non-localised particles occupying volume V at temperature T is given by the equation

$$P = \frac{MkT}{V},$$

$$f_{(s)} = \frac{f'_{(g)} \cdot V}{M} = \frac{f'_{(g)}kT}{P},$$

so that

$$P = kT\frac{f'_{(g)}}{f_{(s)}}. \tag{9.9}$$

Thus the equilibrium vapour pressure of a crystalline solid (that established by the sublimation of a solid in an otherwise empty vessel) is a function only of temperature. It is of interest that, as shown in Section 3.2, the energy levels accessible to a localised particle depend on its molecular volume $\bar{v}$, and, as shown in Section 3.4, that $(\partial \ln f_{(s)}/\partial \bar{v})_T$ is positive. It follows that the subjection of a solid to a pressure *additional to that of its own vapour* leads to a decrease in the value of $f_{(s)}$ and therefore, (from equation 9.9), to an increase in that of the vapour pressure. The above provides a statistical–mechanical foundation for the Poynting equation

$$\left(\frac{\partial \ln P}{\partial P^*}\right)_T = \frac{V_m}{RT},$$

where V_m is the molar volume of the solid and P^* the total pressure applied to it.

One point must be emphasised. The use of equations (9.6) to (9.8) requires that the energy zeros implicit in the chemical potential $\mu_{(s)}$ and in the partition function $f_{(s)}$ are the same as those implicit in the corresponding quantities $\mu_{(g)}$ and $f_{(g)}$. This is just the situation anticipated in Section 3.5, and after reference to that section it should be clear that if we use any of these equations (as we shall do in the next section) to calculate the vapour pressure of a monatomic crystal at temperatures sufficiently high for its behaviour to be represented fairly by the Einstein equations, the appropriate expression for $\mu_{(s)}$ is that obtained from equation (3.45), i.e.

$$\mu_{(s)} = -kT \ln g_{e_0} - \frac{\Delta E^0_{subm}}{L} + 3kT \ln [1 - \exp(-\theta/T)] \qquad (9.10)$$

where ΔE^0 is the molar energy of sublimation at absolute zero.

9.3 The Equilibrium Vapour Pressure of a Monatomic Crystal

We can proceed either from equation (9.6) or from (9.9), but choose the former since expressions for $\mu_{(s)}$ and $\mu_{(g)}$ are readily to hand. Expressing $\mu_{(s)}$ by equation (9.10) and $\mu_{(g)}$ by equation (5.93) i.e.

$$\mu_{(g)} = -kT \ln g_{e_0} - kT \ln \left[\frac{(2\pi m k T)^{\frac{3}{2}} kT}{h^3 P^\dagger} \right] + kT \ln \frac{P}{P^\dagger},$$

substituting into (9.6) and rearranging, we find that[1]

$$\ln \frac{P}{P^\dagger} = -\frac{\Delta E^0_{subm}}{LkT} + 3 \ln [1 - \exp(-\theta/T)] + \ln \left[\frac{(2\pi m k T)^{\frac{3}{2}} kT}{h^3 P^\dagger} \right]. \qquad (9.11)$$

This expression can be improved by substituting for ΔE^0_{subm} an expression for ΔH^T_{subm}, the molar heat of sublimation at temperature T. This is easily obtained. Obviously,

$$\begin{aligned}
\Delta H^T_{subm} &= H^T_{(g)} - H^T_{(s)} \\
&= [H^T_{(g)} - E^0_{(g)}] - [H^T_{(s)} - E^0_{(s)}] + \Delta E^0_{subm}.
\end{aligned}$$

It follows from Section 5.9 that

$$H^T_{(g)} - E^0_{(g)} = \tfrac{5}{2} LkT,$$

and from Section 3.5 that

$$H^T_{(s)} - E^0_{(s)} \approx E^T_{(s)} - E^0_{(s)} = \frac{3Lk\theta}{\exp(\theta/T) - 1}$$

so that

$$\frac{\Delta E^0_{subm}}{LkT} = \frac{\Delta H^T_{subm}}{LkT} - \frac{5}{2} + \frac{3\theta/T}{\exp(\theta/T) - 1}, \qquad (9.12)$$

---

[1] It will be seen that the terms giving the electronic contribution to $\mu_{(s)}$ and $\mu_{(g)}$ cancel, so that (9.11) holds whatever is the state of electronic excitation.

and

$$\ln \frac{P}{P^\dagger} = \frac{5}{2} - \frac{\Delta H^T_{subm}}{LkT} - \frac{3\theta/T}{\exp(\theta/T) - 1}$$

$$+ 3 \ln[1 - \exp(-\theta/T)] + \ln\left[\frac{(2\pi mkT)^{\frac{3}{2}}kT}{h^3 P^\dagger}\right]. \quad (9.13)$$

At temperatures very much greater than θ, $[1 - \exp(-\theta/T)]$ reduces to θ/T and $(\theta/T)[\exp(\theta/T) - 1]^{-1}$ reduces to unity, so that (9.13) can be written in the form

$$\ln \frac{P}{P^\dagger} = -\frac{1}{2} - \frac{\Delta H^T_{subm}}{LkT} + \ln\left[\frac{(2\pi m)^{\frac{3}{2}}k^{\frac{5}{2}}\theta^3}{h^3 P^\dagger T^{\frac{1}{2}}}\right]. \quad (9.14)$$

It is interesting to rearrange this equation to give

$$\ln \frac{P}{P^\dagger} = -\frac{\Delta H^T_{subm}}{LkT} - \frac{1}{2}\ln T/K + \left\{\ln\left[\frac{(2\pi m)^{\frac{3}{2}}k^{\frac{5}{2}}\theta^3}{h^3 P^\dagger K^{\frac{1}{2}}}\right] - \frac{1}{2}\right\} \quad (9.15)$$

where K is here the degree Kelvin and introduced so that the arguments of both logarithmic terms are dimensionless. In thermodynamics it is shown that if the Clausius–Clapeyron equation[1] is integrated taking into account the variation of ΔH_{subm} with temperature, the equation obtained is

$$\ln \frac{P}{P^\dagger} = -\frac{\Delta H^T_{subm}}{RT} + C \ln T/K + I \quad (9.16)$$

where

$$C = \frac{C_{p(g)} - C_{p(s)}}{R}$$

and I is an integration constant the value of which can be obtained only from an experimental value for P. In the case of an atomic species, $C_{p(g)}$ is $5R/2$ and at temperatures very much greater than θ, $C_{(s)}$ equals $3R$ so that the value of C is indeed equal to $-\frac{1}{2}$ as shown in (9.15). Of even greater interest is the fact that (9.15) *predicts* the value of the integration constant I. This is another example of a value which cannot be generated from classical thermodynamics but which is obtainable from the statistical–mechanical treatment.

We now use equation (9.14) to estimate the value of the vapour pressure of silver at 1121 K for comparison with an experimental value. The following information is available:

(a) The vapour pressure of solver is reported[2] to be
  (i) 10^{-5} mm Hg at 1040 K,
  (ii) 10^{-4} mm Hg at 1121 K,
  (iii) 10^{-3} mm Hg at 1209 K.

[1] See Exercise 9.4.
[2] *Handbook of Chemistry and Physics*, 51st Edition, 1970–1, p. D-145.

(b) From (i) and (iii) we find (using the Clausius–Clapeyron equation) that around 1100 K,

$$\frac{\Delta H_{subm}}{2.3026\ R} = 14\ 880\ K.$$

(c) The value for θ for silver varies slightly according to the data from which it is estimated[1] but the mean value is 210 K.

Substituting these quantities into (9.14) we find the theoretical value for the vapour pressure of silver at 1121 K to be 0.0258 Nm^{-2}, whereas (from a(ii) above) the experimental value is 10^{-4} mm Hg i.e. 0.0133 Nm^{-2}. Considering the uncertainty in the value of θ and the even greater uncertainty in the values given in (a), the agreement is not unsatisfactory.

9.4   A Semi-Statistical Method for the Evaluation of the Molar Entropies of all Pure Liquids and Solids

We saw in Chapter 5 that it is possible to obtain, using the methods of the micro-canonical ensemble, extremely accurate values for the molar entropy of any perfect gas, and that it is not difficult to introduce the (usually trivial) correction terms to take into account any departure from perfect gas behaviour. As far as a condensed phase (such as a single-component liquid or solid) is concerned however, no purely statistical method is available. True, an expression for the entropy of any system in terms of Q is given by equation (8.47), but as explained in Section 8.5, it is possible to evaluate Q and so make use of that expression quantitatively only for systems of truly independent particles.

A semi-statistical method is however available, stemming from equation (9.5), (which explains the inclusion of the present section in this chapter) by means of which the molar entropy of *any* single-component condensed phase may be obtained from the statistical value of the entropy of the species as a perfect gas if the value of the equilibrium vapour pressure of the condensed phase at any temperature and the corresponding value for the molar heat of evaporation or sublimation are known. Since the values of heats of evaporation or sublimation can always be estimated using the Clausius–Clapeyron equation (see Exercise 9.5) if values for the vapour pressure at two or more temperatures are known, such values are in fact all that are required, it being assumed, of course, that statistical values for the molar entropies of the corresponding gaseous species are known.

[1] Glasstone, *Textbook of Physical Chemistry* (Macmillan, 1947, p. 422).

For the equilibrium between any species i in a condensed phase c and the corresponding gas we have the relation

$$\mu_{i(c)}^* = \mu_{i(g)}^* \tag{9.17}$$

where the quantities μ^* are the *molar* chemical potentials. If both phases contain only a single species

$$\mu_{i(c)}^* = H_{(c)} - TS_{(c)}$$

and

$$\mu_{i(g)}^* = H_{(g)} - TS_{(g)}$$

where H and S are the molar enthalpies and entropies. Hence

$$S_{(c)} = S_{(g)} - \frac{\Delta H_{c \to g}}{T}, \tag{9.18}$$

where ΔH is the molar heat of transformation $c \to g$, and $S_{(g)}$ is the molar entropy of the gas *at its equilibrium vapour pressure*.

The method is illustrated by calculating the molar entropy of liquid water at 298.15 K. The standard molar entropy of water vapour at that temperature and at one atmosphere pressure is 188.72 JK^{-1}, the molar heat of evaporation 43.928 kJ and its vapour pressure 23.76 mm Hg. Hence,

$$S(H_2O_{(liq)}, 298.15 \text{ K}) = \left[188.72 + R \ln \frac{760}{23.76} \right] - \frac{43\,928}{298.15}$$

$$= 70.20 \text{ JK}^{-1} \text{ mol}^{-1}.$$

9.5 A General Method for the Calculation of the Equilibrium Vapour Pressure of a Condensed Phase

In Section 9.3 a statistical method for calculating the equilibrium vapour pressure of a monatomic solid was described and illustrated. A more general method applicable in principle to any condensed phase containing a single component follows directly from the argument used in the last section, and requires knowledge of the heat of sublimation (or evaporation), the molar entropy of the condensed phase (obtainable presumably by the methods of classical thermodynamics) and the standard molar entropy of the gas (obtainable using the statistical methods of Chapters 4 and 5).

The method will be illustrated by calculating the vapour pressure of graphite at 298.15 K. The following data are available:

(a) the heat of sublimation at this temperature is 715.00 kJ mol^{-1}, (this value is known accurately and is obtained from knowledge of the heat of formation of CO and the bond dissociation energies of CO and O_2),

(b) the experimental value for the entropy of graphite is 5.694 JK^{-1} mol^{-1},

and   (c) and the standard molar entropy of carbon vapour (calculated as a monatomic gas and obtained therefore from equation 5.12) is 157.992 JK^{-1} mol^{-1}.

The value for the equilibrium vapour pressure of graphite expressed in Nm^{-2} is therefore given by the equation

$$S_{(g)} = R \ln \frac{P}{101\,325} + \frac{715\,000}{298.15} + 5.694 = 157.992,$$

from which it follows that

$$P = 4.93 \times 10^{-113} \, \text{Nm}^{-2}.$$

This result is probably as accurate as any quoted in this book, but in order to appreciate the *meaning* of a pressure as low as that obtained it is suggested that the reader carries out Exercise 9.3 *and* studies Exercise 9.6.

EXERCISES

9.1

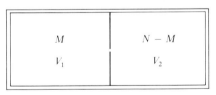

Figure 9.E1

Suppose that N independent non-localised particles are contained in a vessel of total volume V, divided into two parts of volumes V_1 and V_2 by a fixed partition with a small hole in it, as shown in Figure 9.E1.

(a) First, supposing that the system is isolated, show, using the methods of the microcanonical ensemble, that the most probable configuration is that in which the particles are divided into two groups so that M occupy volume V_1 and $N - M$ occupy volume

V_2, M being given by the equation

$$\frac{M}{N-M} = \frac{V_1}{V_2}. \tag{9.E1}$$

(b) Now assume that the system is in equilibrium with a heat bath and use the methods of the canonical ensemble to obtain the same result.
(*Hint*: what distinguishes the particle partition function for particles in one part of the vessel from that for those in the other?)

9.2 The reader may care to see how the condition

$$\mu_{(s)} = \mu_{(g)}$$

may be deduced using the methods of the micro-canonical ensemble. Suppose that a crystalline solid containing N independent localised particles is introduced into a vessel of fixed volume, that the system is then isolated, and that M particles pass into the gas phase.
  Show

(a) that the number of complexions accessible to the gas phase is given by the expression

$$\ln W_{(g)} = M \ln \frac{f_{(g)}}{M} + \beta E_{(g)} + M,$$

and

(b) that the number accessible to the solid phase is given by the expression

$$\ln W_{(s)} = (N - M)\ln f_{(s)} + \beta E_{(s)}.$$

Hence

(c) obtain an expression for $\ln W$ where $W = W_{(g)} \cdot W_{(s)}$.
(d) The equilibrium value for M is that for which

$$\left(\frac{\partial \ln W}{\partial M}\right)_{E,V,N} = 0.$$

Determine this value, remembering that the values of $f_{(g)}$ and $f_{(s)}$ depend on β, and that the equilibrium value of β depends on that of M, (sublimation in an isolated system results in a drop in temperature).

This approach leads to equations (9.8), (9.7) and (9.6) in that order.

9.3 Suppose that a small quantity of graphite is contained in a large vessel of volume V maintained at 298.15 K, and suppose it to be in thermodynamic equilibrium with a single gaseous carbon atom. Use the value given in Section 9.5 for the equilibrium vapour pressure of graphite at this temperature to calculate the value of volume V. What does your result tell you about the chance of finding a single gaseous carbon atom in a vessel volume one litre containing a small quantity of graphite at room temperature?

9.4 One of the most useful relations in physical chemistry is that showing the dependence on temperature of the vapour pressure of a condensed phase containing a single component. In the case of a solid, the relation is

$$\frac{dP}{dT} = \frac{\Delta H_{subm}}{T\Delta V} \tag{9.E2}$$

where ΔV is the difference between the molar volumes of vapour and solid. This equation is exact, but by making two assumptions, one, that the molar volume of the solid is negligible compared to that of its vapour, and the other that the behaviour of the vapour does not differ appreciably from that of a perfect gas, (9.E2) can be expressed less exactly but rather more usefully in the form

$$\frac{d \ln P/P'}{dT} = \frac{\Delta H_{subm}}{RT^2}.$$ (9.E3)

This is usually known as the Clausius–Clapeyron equation. (The symbol P' figuring in (9.E3) is not intended to represent a 'standard' pressure, but the unit in which the vapour pressure is expressed, and is introduced so as to ensure that the argument of the logarithm is dimensionless.)

Show that for a system of independent particles, equation (9.E3) follows directly from equation (9.9).

The corresponding expression to (9.E3) for the equilibrium between a liquid and a gas is obtained by replacing ΔH_{subm} by ΔH_e the molar heat of evaporation. Show that this expression follows from equation (9.5).

9.5   It was stated in Section 9.4 that the value of ΔH_{subm} or that of ΔH_e can be estimated, using the Clausius–Clapeyron equation if values for the vapour pressure at two or more temperatures are known. Substantiate this statement using (9.E3) or the equivalent expression in terms of ΔH_e.

9.6   The standard molar entropy of carbon as a monatomic gas at 298.15 K and 1 atm pressure is 157.9924 JK^{-1}, the molar entropy of diamond at the same temperature is 2.4389 JK^{-1}, the heat of formation of diamond from graphite is 1.8962 kJ mol^{-1}, and the heat of sublimation of graphite is 715.00 kJ mol^{-1}.

(a) Calculate the value for the heat of sublimation of diamond at 298.15 K.
(b) Calculate the equilibrium vapour pressure of diamond at this temperature using the method explained in Section 9.5.
(c) In Table 3.2 the value of the parameter θ for diamond is given as 1364 K. Use this value and that for the heat of sublimation obtained in (a) to estimate the equilibrium vapour pressure of diamond at 298·15 K using equation (9.13).
(d) It will have been found that there is almost a ten-fold difference between the values obtained in (b) and (c), but comparison of these values with that obtained for graphite in Section 9.5 will show that one value is to be preferred to the other. State which, and explain why. Explain also why the method by which the 'discarded' value was obtained is *in this particular case* unsound.

CHAPTER 10

The Adsorption of Gases

1.1 Introduction

A gas is said to be *adsorbed* if its concentration on a surface is greater than that in the gas phase. It was suggested by Langmuir that because of the rapid falling-off of intermolecular forces with distance, it is probable that adsorbed layers are no more than a single molecule thick, (at least at moderately high temperatures and moderately low pressures.) In such a case we would suppose that the fraction of the available surface covered with adsorbed molecules would increase the higher the pressure in the gas phase, and would decrease the higher the temperature.

The first attempt to provide a theoretical foundation for the chemistry of the uni-molecular layer was made by Langmuir. By supposing that the equilibrium state is that at which the rates of condensation onto the surface and evaporation therefrom are the same, he deduced that the pressure required to maintain covered a fraction θ of the total available surface is given by the equation

$$P = \frac{K\theta}{1 - \theta}, \tag{10.1}$$

where the proportionality constant K is a function of temperature. This equation can be rearranged to give

$$\theta = \frac{K^{-1}P}{1 + K^{-1}P}. \tag{10.2}$$

These relationships are known as the *Langmuir adsorption isotherm*. Experimental data have frequently (but not always) been found to be in agreement with them.

Our present task is to obtain an insight into the adsorption process and in particular to provide a statistical–mechanical foundation for the Langmuir equations, so that we may understand the circumstances in which they may be expected to hold. We shall also obtain equations for a type of adsorption which is not in accord with the Langmuir isotherm. The method adopted is similar to that already used in Section 9.1; that is to say we construct an expression for the canonical partition function for a system in which molecules are present both in the gas phase and in an adsorbed layer and determine the equilibrium condition by finding the circumstances in which it is maximised.

Figure 10.1(a) denotes a cross-section of a crystalline surface and a single gas molecule in its vicinity; the lines →— indicating attractive forces between the molecule and the surface atoms. It is reasonable to suppose that the (negative) potential energy of the adsorbed molecule depends on its position relative to the surface atoms, and indeed that if we consider a line x parallel to the grain of the crystal the variation in potential energy would be as shown in Figure 10.1(b), and that the most probable position for an adsorbed molecule would be in one of the potential energy troughs of depth V_0. Furthermore we might suppose that if the kinetic energy of the adsorbed molecule is *less* than V_0 the molecule would be trapped in a potential energy trough (and so be *localised*), whereas if the kinetic energy exceeds V_0, the adsorbed molecules are free to diffuse over the surface, so producing a 'two-dimensional gas' layer. We shall first investigate the case of localised adsorption.

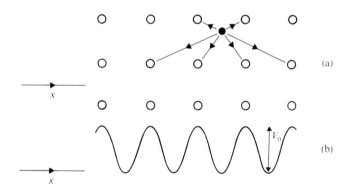

Figure 10.1   Cross-section of a crystal surface and the potential energy of an adsorbed molecule in the plane of the surface.

10.2   Localised (immobile) Adsorption

Suppose that N molecules of a perfect gas are introduced into a vessel of volume V containing a crystalline surface on which are available B equivalent sites (the troughs of Figure 10.1) on each of which a molecule may be adsorbed, and that M molecules are adsorbed leaving $N - M$ in the gas phase. Assume that M is less than B, and suppose that the system is maintained at temperature T.

For any value of M, we may, in accordance with Proposition I of Chapter 8, express the partition function of the system as the product of two subsystem partition functions, one, that of the adsorbed phase and the other that of the gas, i.e.

$$Q_{(M)} = Q_{(a)(M)} \cdot Q_{(g)(N-M)}. \tag{10.3}$$

It follows from (8.84) that

$$Q_{(g)(N-M)} = \frac{f_{(g)}^{(N-M)}}{(N-M)!}. \tag{10.4}$$

The expression for $Q_{(a)(M)}$ is somewhat more complicated. Since each molecule is *located* it is, so to speak, labelled by the position it occupies, so that it is possible, in principle at least, to distinguish any one adsorbed molecule from any other of like kind. It might appear therefore that, in accordance with equation (8,78), $Q_{(a)(M)}$ is given by the expression $f_{(a)}^M$, where $f_{(a)}$ is the partition function for the adsorbed molecule. This however ignores the configurational element. We have B distinguishable sites of which only M are occupied. The number of ways in which M sites can be selected from B is

$$\frac{B!}{M!(B-M)!},$$

and in the expression for $Q_{(a)(M)}$ there must be one term $f_{(a)}^M$ for each distinct geometric configuration. It follows that

$$Q_{(a)(M)} = \frac{B!}{M!(B-M)!} f_{(a)}^M \tag{10.5}$$

so that

$$Q_{(M)} = \frac{B!}{M!(B-M)!} f_{(a)}^M \cdot \frac{f_{(g)}^{(N-M)}}{(N-M)!}. \tag{10.6}$$

Since any number of sites from zero to all may be occupied, $B + 1$ distributional states are possible for the system as a whole, to each of which corresponds a term $Q_{(M)}$ given by equation (10.6).

It follows, in accordance with equation (8.58) that the equilibrium value of M is that leading to the maximum value of $Q_{(M)}$, and so is that value of M satisfying the equation

$$\left(\frac{\partial \ln Q_{(M)}}{\partial M}\right)_{T,V,N} = 0. \tag{10.7}$$

Making use of the Stirling approximation formula, we see that

$$\ln Q_{(M)} = B \ln B - M \ln M - (B - M) \ln (B - M)$$
$$+ M \ln f_{(a)} + (N - M) \ln f_{(g)}$$
$$- (N - M) \ln (N - M) + N - M,$$

so that

$$\left(\frac{\partial \ln Q_{(M)}}{\partial M}\right)_{T,V,N} = - \ln M + \ln (B - M) + \ln f_{(a)}$$
$$- \ln f_{(g)} + \ln (N - M). \tag{10.8}$$

It follows from (10.7) and (10.8) that the equilibrium value for M is that given by the equation

$$\ln \left[\frac{(B - M)(N - M)}{M} \cdot \frac{f_{(a)}}{f_{(g)}}\right] = 0, \tag{10.9}$$

so that

$$\frac{(B - M)(N - M)}{M} \cdot \frac{f_{(a)}}{f_{(g)}} = 1. \tag{10.10}$$

Rearranging, we find that

$$\frac{B - M}{M} \cdot f_{(a)} = \frac{f_{(g)}}{N - M}. \tag{10.11}$$

Since $f_{(g)}$ is proportional to the volume accessible to the molecule, we can write

$$f_{(g)} = f'_{(g)} \cdot V$$

where $f'_{(g)}$ is a function only of temperature, and since for any assembly of $N - M$ independent non-localised particles

$$\frac{V}{N - M} = \frac{kT}{P}$$

where P is the pressure.

$$\frac{B - M}{M} \cdot f_{(a)} = \frac{f'_{(g)}kT}{P},$$

so that

$$P = \frac{M}{B - M}kT\frac{f'_{(g)}}{f_{(a)}}.$$

θ, the fraction of the total number of sites occupied equals M/B, so that we find finally that

$$P = \frac{\theta}{1 - \theta}\left[kT\frac{f'_{(g)}}{f_{(a)}}\right], \tag{10.12}$$

which is obviously equivalent to (10.1), or

$$\theta = \frac{\dfrac{f_{(a)}}{kTf'_{(g)}} \cdot P}{1 + \dfrac{f_{(a)}}{kTf'_{(g)}} \cdot P}, \tag{10.13}$$

which is obviously equivalent to (10.2).

It is clear that the treatment given above not only provides a statistical–mechanical foundation for the Langmuir adsorption isotherm, and identifies the Langmuir proportionality constant K, but implies that the Langmuir equations hold only if the adsorbed molecules are *localised*. We can however, establish the last point without doubt only by analysing a case of *mobile*, non-localised adsorption and showing that in such a case we do *not* arrive at the Langmuir equations, and it is to this task that we turn in the next section. Before doing so we should however draw attention to the fact that the energy zeros implicit in the partition functions $f_{(a)}$ and $f_{(g)}$ figuring in the above equations must be the same. This problem is similar to that considered in Section 3.5 and in the last paragraph of Section 9.2. Reference to these sections should make it clear that the appropriate expression for $f_{(a)}$ is

$$\exp{(e/kT)}f_{\text{int}(a)},$$

where $f_{\text{int}(a)}$ is the partition function for the internal degrees of freedom of the adsorbed molecule (including vibrations relative to its mean position on the site), taking the lowest energy state in the monolayer as the energy zero, and e is the energy required to remove a molecule in its lowest energy state from the monolayer, i.e. $-Le$ is the *molar* energy of adsorption at absolute zero.

10.3   Mobile (non-localised) Adsorption

We now suppose that the adsorbed molecules are free to diffuse over the surface as a two-dimensional gas. We suppose again that M molecules are adsorbed and $N - M$ free, and again write

$$Q_{(M)} = Q_{(a)(M)} \cdot Q_{(g)(N-M)}. \tag{10.14}$$

The expression for $Q_{(g)(N-M)}$ is the same as that given in equation (10.4), but that for $Q_{(a)(M)}$ quite different from that given in (10.5). The essential difference between a localised adsorbed molecule and one which though confined to the surface, is mobile, is that in the case of the latter, the molecule is *not* associated with any particular site, and so is indistinguishable from any of like kind, and for the same reason it is clear that the question of the number of ways in which M sites can be selected from B does not arise. The sub-system partition function $Q_{(a)(M)}$ is therefore $f_{(a)}^M/M!$, from which it follows that

$$Q_{(M)} = \frac{f_{(a)}^M}{M!} \cdot \frac{f_{(g)}^{(N-M)}}{(N-M)!}. \tag{10.15}$$

We have again the situation that since the phases are open, M may assume any value from zero to the maximum permitted by the geometry of the surface, but the most probable (equilibrium) value for M will be that leading to the maximum value of $Q_{(M)}$ and so that satisfying the equation

$$\left(\frac{\partial \ln Q_{(M)}}{\partial M} \right)_{T,V,N} = 0. \tag{10.16}$$

It follows from (10.15) that

$$\ln Q_{(M)} = M \ln f_{(a)} + (N - M) \ln f_{(g)} - M \ln M - (N - M) \ln (N - M) + N,$$

so that

$$\left(\frac{\partial \ln Q_{(M)}}{\partial M} \right)_{T,V,N} = \ln f_{(a)} - \ln f_{(g)} - \ln M + \ln (N - M). \tag{10.17}$$

The equilibrium value of M is therefore that for which

$$\ln \left[\frac{(N - M) f_{(a)}}{M f_{(g)}} \right] = 0,$$

and therefore that given by the equation

$$\frac{f_{(a)}}{M} = \frac{f_{(g)}}{N - M}. \tag{10.18}$$

At this stage, attention is drawn to a second consequence of the fact that the adsorbed molecules are mobile, and moving over the surface as a two-dimensional gas. That is that the partition function $f_{(a)}$ will contain as a factor a two-dimensional translational partition function. Reference to the equations of Section 5.1 will show this factor to be $(2\pi mkT/h^2)\,\mathscr{A}$, where $\mathscr{A}$ is the area of the surface. In other words $f_{(a)}$ can be expressed as $f'_{(a)}\mathscr{A}$ where $f'_{(a)}$, equal to $(2\pi mkT/h^2)\cdot f_{int}$ is a function only of temperature.

On the other hand $f_{(g)}$ is, as before, proportional to the volume accessible to the free molecules so that $f_{(g)} = f'_{(g)}V$, and, as before

$$\frac{N - M}{V} = \frac{P}{kT},$$

so that

$$\frac{M}{\mathscr{A}} = \frac{f'_{(a)}}{kTf'_{(g)}} \cdot P. \tag{10.19}$$

It is debatable whether we can legitimately speak of the number of available *sites* in the case of non-localised adsorption, because the whole area $\mathscr{A}$ is available, but legitimate or not it is useful to do so in order to produce an equation analogous to (10.13). We therefore write $\mathscr{A} = \alpha B$ where α is a proportionality constant and defining θ as M/B as before, find that

$$\theta = \alpha\frac{f'_{(a)}}{kTf'_{(g)}} \cdot P. \tag{10.20}$$

This equation is recognisably *not* the Langmuir isotherm, so that we conclude that the Langmuir equations apply only to the case of localised adsorption.

Equation (10.20) shows that the amount of gas adsorbed onto a given surface in the case of mobile adsorption is linearly proportional to the pressure of the gas phase in equilibrium with it. Such behaviour has been observed, but more frequently than not, experimental data show that the amount of gas adsorbed is proportional to P^n where n deviates slightly from unity. We do not need to go far to find the reason for this slight divergence from equation (10.20). This equation was deduced on the assumption that the molecules are *independent* in both the gas phase and the adsorbed layer. Although this assumption is entirely reasonable in the gas phase (the divergence of the behaviour of a real gas at relatively low pressures from that of a perfect gas being very small indeed) we would expect intermolecular forces to come into play in the adsorbed phase because of the close proximity of one adsorbed molecule to another. The reason why this is apparently of negligible importance in the case of localised adsorption is presumably due to the fact that each adsorbed molecule is necessarily separated from its neighbours by the potential energy barrier V_0.

We finally draw attention to the fact that we return to the subject of the adsorption of gases in Chapter 13, where, using the relations of the grand canonical ensemble, we obtain the Langmuir equations rather more directly than they were obtained in Section 10.2, and in addition, consider two interesting cases, one in which molecules of different components are in competition for available sites, and the other the case in which atomic species in the monolayer are in equilibrium with diatomic molecules in the gas phase.

EXERCISES

10.1 The expressions for the Langmuir isotherm may be reached by another method, that of equating the expression for the chemical potential of a molecule in the gas phase with that of a molecule in the adsorbed layer.

(a) Show, following equation (10.6), that for any arbitrary value of M, the Helmholtz function A is given by the equation

$$A = -kT \ln \left[\frac{B!}{M!(B - M)!} f_{(a)}^M \cdot \frac{f_{(g)}^{(N-M)}}{(N - M)!} \right]. \tag{10.E1}$$

(b) Observe that this expression is separable into two parts, one of which is the Helmholtz function for the gas, and the other that for the adsorbed phase, i.e.

$$A_{(g)} = -kT \ln \left[\frac{f_{(g)}^{(N-M)}}{(N - M)!} \right], \tag{10.E2}$$

and

$$A_{(a)} = -kT \ln \left[\frac{B!}{M!(B - M)!} f_{(a)}^M \right]. \tag{10.E3}$$

(c) Obtain the values of $\mu_{(g)}$ and $\mu_{(a)}$ from the relations

$$\mu_{(g)} = \frac{\partial A_{(g)}}{\partial (N - M)} \quad \text{and} \quad \mu_{(a)} = \frac{\partial A_{(a)}}{\partial M},$$

and equate.

10.2 Carry out a similar exercise to 10.1 for a system in which the adsorption is mobile.

10.3 The reader may care to see how the expressions for the Langmuir isotherm may be reached using the (much more laborious) methods of the micro-canonical ensemble.

Suppose that N molecules of a perfect gas are introduced into a vessel of volume V, containing a surface on which are available B equivalent sites, that the system is isolated, and that M molecules are adsorbed, leaving $N - M$ in the gas phase. Denote the characteristics of the molecules in the adsorbed phase by the symbols ϵ_i, g_i and n_i, and those of the molecules in the gas phase by the symbols ε_j, ϖ_j and m_j.

(a) Show that for any value of M, the number of complexions accessible to the system is given by the equation

$$W_{(M)} = \frac{B!}{M!(B - M)!} M! \prod_i \frac{g_i^{n_i}}{n_i!} \cdot \prod_j \frac{\varpi_j^{m_j}}{m_j!},$$

where the population numbers n_i and m_j satisfy the equations

$$\sum_i n_i + \sum_j m_j = N$$

and

$$\sum_i n_i \epsilon_i + \sum_j m_j \varepsilon_j = E.$$

(b) Obtain expressions for the most probable values of n_i and m_j in the usual way, and so obtain an expression for ln $W_{(M)}$ in terms of M, β, $f_{(a)}$ and $f_{(g)}$.

(c) The equilibrium value for M is that for which ln $W_{(M)}$ is a maximum, i.e. that value of M satisfying the condition

$$\left(\frac{\partial \ln W_{(M)}}{\partial M}\right)_{E,V,N} = 0.$$

Solve this equation, taking into account the fact that β, $f_{(a)}$ and $f_{(g)}$ are all functions of M. Explain why this is so.

CHAPTER 11

Liquid Mixtures

In this chapter we present a short description of the most important features of liquid mixtures, and in later sections seek a statistical–mechanical explanation for these features. This topic is a valuable one from the instructional viewpoint, because it serves as an excellent illustration of the fact that though we are unable to evaluate Q for a system of interacting particles, we are able, in some instances, to factorise Q into supposedly independent partition functions for different degrees of freedom (Postulate II of Chapter 8), and to evaluate some of these factors, and so able to offer partial solutions to problems quite beyond the resources of the mechanics described in the first part of the book.

We shall concern ourselves only with the behaviour of *binary* liquid mixtures, i.e. mixtures of two molecular species A and B, but the extension of all equations to cover multi-component mixtures is obvious. We shall denote the amounts of each component in the mixture by the number of molecules of each, denoting the number of molecules of A and B by the symbols N and M respectively, and shall denote the composition of the mixture by the single parameter x, this representing the mole fraction of component A, i.e.

$$x = \frac{N}{N + M},$$

the mole fraction of component B, $(1 - x)$, being given by the equation

$$1 - x = \frac{M}{N + M}.$$

When referring to the chemical potential of A or B we mean the molecular chemical potential. The equations of the first section thus differ slightly from those presented in most texts on classical thermodynamics in the sense that there the amounts of each component are usually expressed in terms of moles, and the chemical potentials employed are the *molar* not *molecular* quantities. We shall first consider mixtures of components which are miscible over the entire composition range from $x = 0$ to $x = 1$, but later in the chapter we shall uncover the conditions which appear to operate in the case of limited miscibility.

11.1   The Physical Chemistry of Binary Liquid Mixtures

Much of our knowledge of the behaviour of liquid mixtures stems from measurements of the partial vapour pressures of their components, and from the dependence of such quantities on the composition of the liquid phase. Experiment shows that at any fixed temperature, the partial vapour pressures of components of mixtures miscible over the whole composition range approximate to one of the three types of behaviour represented by Figure 11.1(a), (b) and (c).

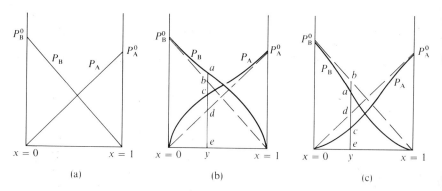

Figure 11.1   The dependence of the partial vapour pressures of binary liquid mixtures at constant temperature on composition.

The partial vapour pressures of components of mixtures represented by (a) are, within the limits of experimental error, linearly proportional to the appropriate mole fractions, so providing the relations

$$P_A = xP_A^0, \qquad P_B = (1 - x)P_B^0, \tag{11.1}$$

where P_A^0 and P_B^0 are respectively the vapour pressures of A and B as pure liquids at the same temperature. The relations shown in (11.1) are known as Raoult's law, and mixtures which obey Raoult's law over the entire composition range are known as *perfect mixtures*, (in some texts as *ideal* mixtures or *ideal solutions*.) By equating the chemical potential of a component of a liquid mixture with that of the component of the vapour with which it is in equilibrium, and making use of equation (6.39), we see that

$$\mu_{A(liq)} = \mu_{A(g)} = \mu_A^\dagger + kT \ln \frac{P_A}{P^\dagger}$$

$$= \mu_A^\dagger + kT \ln \frac{P_A^0}{P^\dagger} + kT \ln x$$

$$= \mu_A^\circ + kT \ln x, \tag{11.2}$$

where μ_A° is defined by the relation

$$\mu_A^\circ = \mu_A^\dagger + kT \ln \frac{P_A^0}{P^\dagger}.$$

Since equation (11.2) holds over the entire composition range including the case of $x = 1$, it is evident that μ_A° is the chemical potential of A as a pure liquid at temperature T. A similar equation

$$\mu_{B(liq)} = \mu_B^\circ + kT \ln (1 - x) \tag{11.3}$$

obviously holds for component B. Equations (11.2) and (11.3) are equivalent in every sense to relations (11.1), and may be chosen equally well, to define the perfect mixture. In obtaining these equations, we have assumed that the vapour phase obeys the equations of a perfect gas. Any error resulting from this rarely exceeds the uncertainty with which values of partial vapour pressures are known, and in any event, the equations may be made formally correct by substituting fugacities for partial pressures as described in most thermodynamic texts.

It follows from (11.2) and (11.3) that the change in Gibbs function resulting from the formation (at constant temperature and pressure) of a perfect mixture from a pure liquid containing N molecules of A and a second pure liquid containing M molecules of B is given by the equation

$$\Delta G_m = NkT \ln x + MkT \ln (1 - x), \tag{11.4}$$

the corresponding entropy change by the equation

$$\Delta S_m = - \left(\frac{\partial (\Delta G_m)}{\partial T} \right)_P = - Nk \ln x - Mk \ln (1 - x), \tag{11.5}$$

and the heat of mixing by the equation

$$\Delta H_m = \Delta G_m + T \Delta S_m = 0. \tag{11.6}$$

Perfect liquid mixtures are rare, and indeed the equations of the perfect mixture are probably obeyed *precisely* only by mixtures of molecules which differ only by isotopic substitution, (e.g. C_6H_6/C_6D_6), but a few mixtures of molecules of very similar structures, one example being ethylene bromide/ propylene bromide, and another benzene/bromobenzene show behaviour very close indeed to that of the perfect mixture, and if we allow about one per cent deviation we find several more: two examples being benzene/toluene, and methyl chloride/chloroform.

The great majority of mixtures exhibit the behaviour illustrated by (b) and (c), and such mixtures are known as *imperfect*. Those illustrated by (b), in which for all compositions the partial vapour pressures are never less than those which would be displayed were Raoult's law obeyed, are said to exhibit *positive* deviations from Raoult's law, and those illustrated by (b) are said to exhibit *negative* deviations.

The properties of imperfect mixtures are correlated by replacing the Raoult law equations (11.1) by the equations

$$P_A = \gamma_A x P_A^0, \qquad P_B = \gamma_B(1 - x)P_B^0, \qquad (11.7)$$

where γ_A and γ_B are dimensionless quantities the values of which depend on the temperature and composition of the mixture. For a mixture of composition $x = y$ as shown in Figure 11.1(b) and (c), it follows from (11.7) that γ_A is given by the ratio of the length of the line ce to that of the line de, and γ_B by the ratio of the length of the line ae to that of the line be. It is clear, first, that the divergence of γ_A and γ_B from unity measures the degree of departure of the mixture (at the chosen composition) from Raoult's law, so that it is appropriate that these quantities are called (*Raoult*) *activity coefficients*, and second, that in the case of mixtures exhibiting positive deviations the activity coefficients are never less than unity over the entire composition range, and that in the case of negative deviations the activity coefficients never exceed unity.

An important feature of all imperfect mixtures is that in the region in which one component predominates so greatly that its mole fraction approaches unity, its vapour pressure curve becomes asymptotic to the Raoult line, so that as $x \to 1$, $\gamma_A \to 1$, and as $x \to 0$, $\gamma_B \to 1$.

By precisely the same method as was used above in the case of the perfect mixture, it may be shown that in the case of imperfect mixtures, the relations given in (11.7) are equivalent to the expressions

$$\mu_{A(liq)} = \mu_A^\circ + kT \ln \gamma_A x, \qquad (11.8)$$

and

$$\mu_{B(liq)} = \mu_B^\circ + kT \ln \gamma_B(1 - x), \qquad (11.9)$$

so that the thermodynamic quantities of mixing are given by the equations

$$\Delta G_m = NkT \ln \gamma_A x + MkT \ln \gamma_B (1 - x), \qquad (11.10)$$

$$\Delta S_m = -Nk \ln \gamma_A x - NkT \frac{\partial \ln \gamma_A}{\partial T}$$

$$- Mk \ln \gamma_B (1 - x) - MkT \frac{\partial \ln \gamma_B}{\partial T}, \qquad (11.11)$$

and

$$\Delta H_m = -NkT^2 \frac{\partial \ln \gamma_A}{\partial T} - MkT^2 \frac{\partial \ln \gamma_B}{\partial T}. \qquad (11.12)$$

The experimental measurements most usually made on mixtures are those in which heats of mixing are obtained directly, and those of the partial pressures, which yield experimental values for γ_A, γ_B and ΔG_m, it following from equations (11.1), (11.4), (11.7) and (11.10) that

$$\Delta G_m = NkT \ln \frac{P_A}{P_A^0} + MkT \ln \frac{P_B}{P_B^0}. \qquad (11.13)$$

Experimental values for ΔS_m are obtained from these using the relation

$$T \Delta S_m = \Delta H_m - \Delta G_m.$$

It is demonstrated in most thermodynamics texts that the values of γ_A and γ_B play a vital rôle in determining the behaviour of the mixture of the particular composition to which the values refer. Two examples suffice. If the values of γ_A and γ_B are very close to unity over the whole composition range, the components of the mixture may be separated completely by fractional distillation. If however they depart appreciably from unity, the mixture exhibits *azeotropy*, (an azeotropic mixture is one in which for one value of x the composition of the liquid phase is the same as that of the vapour with which it is in equilibrium,) and complete separation by fractional distillation is not possible. A second example is that if the mixture shows *positive* deviations, and the activity coefficients exceed particular critical values, reduction in temperature sometimes leads to separation of the hitherto homogeneous mixture into two immiscible layers, one a saturated solution of A in B, and the other a saturated solution of B in A.

Of obvious importance is the question of the dependence of the values of activity coefficients on the composition of the mixture. Classical thermodynamics gives no information on this other than emphasising that each *pair* of activity coefficients is subject to the Gibbs–Duhem equation, [1,2] which when applied to binary liquid mixtures at constant temperature and pressure takes the form

$$x\left(\frac{\partial \ln \gamma_A}{\partial x}\right)_{T,P} - (1-x)\left(\frac{\partial \ln \gamma_B}{\partial(1-x)}\right)_{T,P} = 0, \qquad (11.14)$$

so that the shape of the partial pressure curve of one component determines that of the other, and the earliest work in this field, (associated with the names Margules, von Laar, Porter and others), was to establish *empirical* relations between γ_A, γ_B and x.

The first step in this exercise was made by Margules (1895) who demonstrated that the partial vapour pressures of components of all mixtures can be represented by such equations as

$$\ln \frac{P_A}{P_A^0 x} = b(1-x)^2 + c(1-x)^3 + d(1-x)^4 + \cdots$$

$$(11.15)$$

$$\ln \frac{P_B}{P_B^0(1-x)} = b'x^2 + c'x^3 + d'x^4 + \cdots$$

the coefficients $b, c, d, \ldots, b', c', d', \ldots$ being of course interrelated through the Gibbs–Duhem equation (see Exercise 11.1). This observation was made prior to the introduction of activity coefficients, but it follows from (11.7) that the Margules equations are equivalent to the relations

$$\ln \gamma_A = b(1-x)^2 + c(1-x)^3 + d(1-x)^4 + \cdots$$

$$(11.16)$$

$$\ln \gamma_B = b'x^2 + c'x^3 + d'x^4 + \cdots .$$

Such information as was then available indicated that the coefficients c and c' are small compared with b and b', and that the coefficients d and d' are smaller still.

In 1920 Porter[3] showed that the partial vapour pressures of acetone/diethyl ether mixtures at 30°C calculated from the equations

$$\ln \frac{P_A}{P_A^0 x} = 0.74(1-x)^2$$

$$(11.17)$$

$$\ln \frac{P_B}{P_B^0(1-x)} = 0.74x^2$$

[1] It is from this equation that it may be shown that if one component exhibits positive deviations from Raoult's law over the entire composition range, so must the second, and if one exhibits negative deviations over the whole range so must the second.

[2] The Gibbs–Duhem equation is derived in Appendix 1.

[3] Porter, *Trans. Faraday Soc.* **16**, 336 (1920).

agree well with the experimental values. These equations are of course equivalent to the relations

$$\ln \gamma_A = b(1 - x)^2$$

$$\ln \gamma_B = bx^2.$$

(11.18)

Since that time many imperfect mixtures showing *positive* deviations from Raoult's law have been found to conform to relations (11.18). Imperfect mixtures which show negative deviations and conform to relations (11.18) are rare, but known, one example being chloroform/acetone.

It is evident that we may regard equations (11.16) as the general empirical relation between γ_A, γ_B and x, mixtures which conform to (11.18) as a restricted class, (for convenience known usually as *regular mixtures*,) for which c, c', d and $d', \ldots$ are zero and $b = b'$, and *perfect mixtures* as a still further restricted class for which b, b', c, c', d and d' are all zero.

We are concerned here only with liquid mixtures, but it is pertinent to point out that the behaviour of *solid solutions* is very similar indeed to that of liquid mixtures, and that examples of solid solutions may be found which are 'perfect' or 'regular' in the sense that the terms are here used.

The object of the following sections is to use the methods of statistical mechanics to provide insight into the conditions operating at the molecular level such as would lead to the various types of behaviour described above.

11.2 Statistical–Mechanical Theories of Mixing Processes: A Preliminary Discussion

A complete statistical–mechanical theory of liquids and liquid mixtures would be one which set up models from which could be obtained expressions for the thermodynamic properties of each of any two pure liquids A and B and those of the mixture in terms of the characteristics of the molecules concerned. Such a theory is quite beyond the resources of statistical mechanics at the present time. A satisfactory statistical–mechanical theory of pure liquids is still awaited, and the fact that moderately satisfactory theories of the *mixing process* have been achieved is due solely to the fact that the objective has been severely restricted: no attempt being made to produce expressions which lead to absolute values for the properties of either the pure liquids or the mixture, but only to produce expressions for the *changes* in those quantities which are associated with the mixing process in terms of parameters to which may be attributed a physical significance, but the values of which are determined only by experiment.

Unambitious as this programme appears, we shall find that it provides considerable insight into the thermodynamics of mixing processes, and it is perhaps rather surprising, considering the assumptions and approximations made at each stage of the argument that the final expressions obtained correspond as closely as they do to the relations operating in the case of many real mixtures.

We must make it clear also that this particular field is one to which a great deal of attention has been paid for many years, and one in which work is still in progress, and we shall provide here only a very simplified description of some of the work which has been done. Indeed, we shall consider the foundations of only one line of attack. Most attempts to set up a statistical mechanics of liquid mixtures fall into one of three groups:

(i) those theories which are based on what are usually called *quasi-crystalline lattice models* of liquids and liquid mixtures;

(ii) those based on a more sophisticated approach due mainly to Prigogine and his co-workers, and described very fully in his monograph *The Molecular Theory of Solutions*, (North-Holland Publishing Company, 1957); and

(iii) those approaches, due originally to Longuet–Higgins[1] based on the *principle of corresponding states*.

We shall concern ourselves only with (i), because for all their deficiencies, the theories based on the quasi-crystalline lattice models show very clearly the conditions which must operate at the molecular level in order to provide the equations of the perfect mixture, and how relaxation of some conditions leads to the behaviour typified by relations (11.18) and (11.16).

We now discuss the principal assumptions we make at the start of our attack.

Assumption 1

The first assumption is, as implied above, that both pure liquids and liquid mixtures have quasi-crystalline lattice structures in which each molecule is surrounded by a definite number of nearest neighbours, and in which each molecule is distinguished by the site it occupies.

This assumption is necessary because, as was stated earlier, no satisfactory statistical–mechanical theory of the liquid state is yet available. The notion that this assumption is perhaps rather more realistic than it might appear at first sight, is supported, first by the fact that the density of any liquid is much the same as that of the corresponding crystalline solid, so that it is evident that the molecules in a liquid must be packed together in much the same way as they are in a solid, and so subject to much the same inter-molecular forces, and secondly by X-ray analysis which shows that a liquid may be

---

[1] Longuet–Higgins, *Proc. Roy. Soc.* **205A**, 247 (1951).

regarded as a rather loose structure in which the molecules are arranged in a way not unlike that of a crystal except that it lacks the almost complete order of the latter. It is true of course, that this assumption denies to liquids one of their most important characteristics, that of fluidity, and indeed, if we were to attempt to use such a model to obtain absolute values for any of the properties of a liquid or liquid mixture, such attempts would be bound to fail. This however, is not our aim: as explained earlier, we make no attempt to calculate the absolute values of the properties of pure liquids or liquid mixtures, but only the *changes* in such properties when mixtures are formed from their pure components, so that an unrealistic model of the liquid state may well suffice if it happens that its inadequacy to describe the mixture is much the same as its inadequacy to describe the pure components.

Assumption 2

The second assumption is that the molecules of both species A and B are of approximately the same size and shape, and that the lattice structures of both the pure liquids and the mixture are the same. This assumption implies that any one molecule is surrounded by the same number of nearest neighbours whether it be in the pure liquid or in the mixture. It also implies that the volume change during the mixing process is negligible. This last implication carries with it a not unimportant benefit when comparing the statistical–mechanical expressions produced later with the thermodynamic expressions given in Section 11.1. The latter refer to mixing processes carried out at constant temperature and pressure, and the natural quantities by which such processes are described are ΔG, the change in Gibbs function, and ΔH, the *heat* of mixing. On the other hand, the statistical–mechanical machinery we constructed in Chapter 8 leads directly to changes in properties resulting from processes carried out at constant temperature and at constant volume, and the natural quantities by which such processes are described are ΔA, the change in Helmholtz function, and ΔE, the *energy* of mixing. A consequence of the assumption that the volume of the mixture is the same as the sum of the volumes of the pure liquids from which it is formed is that the quantities ΔG_m and ΔH_m figuring in Section 11.1 may be compared directly with the quantities ΔA_m and ΔE_m derived in later sections.

Assumption 3

The third assumption is that the forces operating between molecules other than those which are nearest neighbours are negligible. This assumption is perhaps less sweeping than it appears at first sight, but we shall make no attempt to justify it.

Assumption 4

The fourth assumption is that the molecules of both species A and B are non-polar. The reason for this is simple. Polar molecules would tend to assume relatively highly-ordered configurations commensurate with the minimum electrostatic energy. We choose not to concern ourselves with such complications, but to consider mixtures in which the distribution of molecules is completely random.

Assumption 5

It will be realised that both pure liquids and liquid mixtures are cooperative systems, systems of interacting particles. Such systems can therefore be described only by the equations of the canonical ensemble. The most important quantity with which we shall be concerned is the change in Helmholtz function resulting from a process in which liquids A and B mix at constant temperature, and it follows from the relations established in Chapter 8 that

$$\Delta A_m = A_{\text{mixture}} - (A_A + A_B)$$

$$= -kT \ln \frac{Q}{Q_A \cdot Q_B}, \tag{11.19}$$

where Q_A is the partition function for liquid A before mixing, Q_B that of liquid B and Q that of the mixture. Since the pure liquids and the mixture are systems of interacting particles we have no means of determining any of the three partition functions, but this notwithstanding, it is possible after making certain assumptions, to obtain an approximate expression for the quotient $Q/Q_A \cdot Q_B$, and so an approximate expression for ΔA_m.

Let us consider the partition function Q_A for a pure liquid containing N molecules of species A.

We first suppose[1] that the degrees of freedom relating to the position and motion of the centre of mass of each molecule on or about its lattice site are separable from its internal degrees of freedom (its rotational activity and the vibrational activity of the atoms within the molecule,) so that the energy of the system may be resolved into independent components,

$$E_A = E_{(t)A} + E_{(int)A}. \tag{11.20}$$

Following Proposition II of Chapter 8, we therefore write

$$Q_A = Q_{(t)A} \cdot Q_{(int)A}. \tag{11.21}$$

We further suppose that $E_{(t)A}$ may be resolved into two parts, one of which is that related to the molecular vibrations of each molecule about its lattice

[1] We here follow the approach used in Guggenheim's *Mixtures*, page 15 (Oxford, 1952).

point, and the other, which we call the *configurational* energy, that (other than the energy $E_{(int)}$) which the system would possess were the centre of mass of each molecule at rest on its lattice point, so that

$$E_{(t)A} = E_{(c)A} + E_{(vib)A}. \tag{11.22}$$

This separation implies that $Q_{(t)A}$ may be factorised as shown:

$$Q_{(t)A} = Q_{(c)A} \cdot Q_{(vib)A}, \tag{11.23}$$

or, making use of (11.21), that

$$Q_A = Q_{(c)A} \cdot Q_{(vib)A} \cdot Q_{(int)A}. \tag{11.24}$$

We suppose that similar resolutions and factorisations are permissible for liquid B and for the mixture.

The final assumption is that the internal and vibrational degrees of freedom of a molecule A are unaffected by the mixing process: in other words, that the contributions of these terms to the partition function Q_A relating to a liquid containing only N molecules of A are the same as those to the partition function Q relating to a mixture containing N molecules of A in the presence of M molecules of B, and that the same applies to the contributions due to component B. It follows that the only changes which occur during the mixing process are changes in the configurational energy and changes in the configurational partition functions, so that (11.19) may be replaced by the equation

$$\Delta A_m = -kT \ln \frac{Q_{(c)}}{Q_{(c)A} \cdot Q_{(c)B}}, \tag{11.25}$$

where $Q_{(c)}$ is the configurational partition function for the mixture. All that remains is to show how expressions for $Q_{(c)}$, $Q_{(c)A}$ and $Q_{(c)B}$ may be framed. For this we require expressions for the configurational energy.

Before proceeding to this it is perhaps as well to emphasise that the final assumptions made above are by no means unrelated to assumptions 2 and 4 made earlier. If molecules A and B were of different sizes and shapes the mixing process would result in distortions in the lattice and to changes in the quantities $Q_{(vib)A}$ and $Q_{(vib)B}$, and if the molecules were polar the mixing process would lead to changes in both these and the quantities $Q_{(int)A}$ and $Q_{(int)B}$.

11.3 Configurational Energies

Suppose that in a liquid containing N molecules of A, each molecule is surrounded by z nearest neighbours, and that (following assumption 3) the forces of attraction between any two molecules which are not nearest neighbours can be ignored. Denote the work required to extract one molecule from its cluster of z nearest neighbours by the term $-z\epsilon_{AA}$. Note that the quantity $-z\epsilon_{AA}$ is positive and therefore the term ϵ_{AA} negative. We can represent the extraction process by the scheme

(i)

$$+ \text{(A)}. \qquad w(i) = -z\epsilon_{AA}.$$

The total number of nearest neighbour pairs in the liquid is $\frac{1}{2}zN$, so that the configurational energy of the liquid A *relative to that of N molecules at infinite separation* is therefore given by the equation

$$E_{(c)A} = \tfrac{1}{2}zN\epsilon_{AA}. \tag{11.26}$$

This quantity is of course negative.

We now carry out the same operation for a liquid containing M molecules of B, supposing (assumption 2) that here again each molecule is surrounded by z nearest neighbours. Denoting the work required to extract one molecule from its z nearest neighbours by the term $-z\epsilon_{BB}$, as illustrated by the scheme

(ii)

$$+ \text{(B)}, \qquad w(ii) = -z\epsilon_{BB},$$

we see that the configurational energy of liquid B *relative to that of M molecules at infinite separation* is given by the equation

$$E_{(c)B} = \tfrac{1}{2}zM\epsilon_{BB}. \tag{11.27}$$

We now suppose that the molecule A extracted in (i) is placed into the cavity in liquid B, and that the molecule B extracted in (ii) is placed into the cavity in liquid A. In effect, we are now forming $2z(AB)$ nearest neighbour pairs, and we denote the work required to do so by the term $2z\epsilon_{AB}$. (This quantity of work is negative.) It follows that the work required to replace one molecule in liquid A by a molecule B, and one molecule in liquid B by a molecule A (i.e. separating $z(AA)$ pairs and $z(BB)$ pairs and forming $2z(AB)$ pairs is

$$2z[\epsilon_{AB} - \tfrac{1}{2}(\epsilon_{AA} + \epsilon_{BB})],$$

and that for a process resulting in the formation of zX nearest-neighbour unlike pairs is

$$zX[\epsilon_{AB} - \tfrac{1}{2}(\epsilon_{AA} + \epsilon_{BB})]. \tag{11.28}$$

We now define the parameter w' by the equation

$$w' = \epsilon_{AB} - \tfrac{1}{2}(\epsilon_{AA} + \epsilon_{BB}), \tag{11.29}$$

so that the quantity given in (11.28) may be expressed more economically as

$$zXw'.$$

It follows that the configurational energy of a mixture of N molecules of A and M molecules of B *relative to that of all molecules at infinite separation* is, supposing the mixture to contain zX nearest neighbour unlike pairs, given by the equation

$$E_{(c)} = \tfrac{1}{2}zN_{AA} + \tfrac{1}{2}zM_{BB} + zXw'. \tag{11.30}$$

The configurational partition function for liquid A is by definition

$$Q_{(c)A} = \sum W_{(c)A} \exp\left(-E_{(c)A}/kT\right),$$

the summation being carried out over all possible values of $E_{(c)A}$ accessible to the system maintained at temperature T. It follows however from the study of fluctuations made in Section 8.3 that all terms in the sum are negligible other than those for energies indistinguishable from the most probable value for $E_{(c)A}$, and further, that the logarithm of $Q_{(c)A}$ is indistinguishable from that of its maximum term, so making use of (11.26), we see that

$$\ln Q_{(c)A} = \ln W_{(c)A} - \frac{E_{(c)A}}{kT}$$

$$= \ln W_{(c)A} - \frac{zN\epsilon_{AA}}{2kT}. \tag{11.31}$$

Similarly, making use of (11.27) and (11.30) we obtain the equations

$$\ln Q_{(c)B} = \ln W_{(c)B} - \frac{zM\epsilon_{BB}}{2kT} \tag{11.32}$$

and

$$\ln Q_{(c)} = \ln W_{(c)} - \left[\frac{zN\epsilon_{AA}}{2kT} + \frac{zM\epsilon_{BB}}{2kT} + \frac{zXw'}{kT}\right]. \tag{11.33}$$

Substituting into (11.25) we find that

$$\Delta A_m = -kT \ln \frac{W_{(c)}}{W_{(c)A} \cdot W_{(c)B}} + zXw'. \tag{11.34}$$

This is the fundamental equation for the mixing process at constant temperature for a system subject to the conditions assumed in Section 11.2.

11.4   The Statistical Mechanics of Perfect Mixtures

We now postulate that the sufficient condition for a mixture to be perfect is that the quantity w' defined by equation (11.29) is zero at all temperatures of interest. Before substituting this value into (11.34) and exploring the consequences, we must make the point that it is reasonable to put w' equal to zero only (i) if the molecules A and B *are* of approximately the same size and shape so that a molecule B may replace a molecule A on a lattice site without causing strain (and so increasing the configurational energy), and (ii) if molecules A and B *are* surrounded by the same number of nearest neighbours in both the pure liquids and in the mixture. Both of these conditions were of course imposed in Section 11.2 as assumption 2, but we here emphasise that assumption 2 is prerequisite to the condition $w' = 0$.

We can now proceed without further ado. It follows from (11.34) that for the perfect mixture

$$\Delta A_m = -kT \ln \frac{W_{(c)}}{W_{(c)_A} \cdot W_{(c)_B}}. \tag{11.35}$$

The terms $W_{(c)}$, $W_{(c)_A}$ and $W_{(c)_B}$ do not include the number of ways in which the total energy of the system is distributed among the constituent molecules, nor the degeneracies of each energy level accessible to each molecule. Such numbers form part of the quantities $Q_{(vib)_A}$, $Q_{(int)_A}$, $Q_{(vib)_B}$ and $Q_{(int)_B}$ and are supposedly unchanged during the mixing process, (assumption 5). $W_{(c)}$ is simply the number of distinguishable geometric arrangements of N identical molecules A and M identical molecules B on $N + M$ equivalent 'labelled' sites. This number is

$$\frac{(N + M)!}{N!M!}.$$

There is only one distinguishable arrangement of N identical molecules of A on N 'labelled' sites, so that $W_{(c)_A}$ is unity, as is $W_{(c)_B}$. It follows that

$$\Delta A_m = -kT \ln \frac{(N + M)!}{N! M!},$$

so, making use of the Stirling approximation formula

$$\Delta A_m = -(N + M)kT \ln (N + M) + NkT \ln N + MkT \ln M$$

$$= NkT \ln \frac{N}{N + M} + MkT \ln \frac{M}{N + M}$$

$$= NkT \ln x + MkT \ln (1 - x). \tag{11.36}$$

It follows that

$$\Delta S_m = - \left(\frac{\partial(\Delta A_m)}{\partial T}\right)_V = -Nk \ln x - Mk \ln (1 - x), \qquad (11.37)$$

and

$$\Delta E_m = \Delta A_m + T \Delta S_m = 0. \qquad (11.38)$$

It will be recalled that it was pointed out in Section 11.2, that as a consequence of the implication that the mixing process brings about no volume change, the quantities ΔG_m and ΔH_m figuring in Section 11.1 may be compared directly with the statistical–mechanical expressions for ΔA_m and ΔE_m. Bearing this in mind, it is clear that equations (11.36) to (11.38) are equivalent entirely to equations (11.4) to (11.6).

The Gibbs function for the pure liquid A is given by the equation

$$G_A = N\mu_A^\circ,$$

where μ_A° is the chemical potential of A as a pure liquid, and since (as pointed out first in Section 3.4) we can equate the Gibbs function of any condensed system with its Helmholtz function without undue error, we may write

$$A_A \simeq N\mu_A^\circ$$

and

$$A_B \simeq M\mu_B^\circ.$$

It follows that the Helmholtz function for the perfect mixture is given by the expressions

$$A_{\text{mixture}} = A_A + A_B + \Delta A_m$$
$$= N\mu_A^\circ + M\mu_B^\circ + NkT \ln x + MkT \ln (1 - x), \qquad (11.39)$$

so that the chemical potentials of A and B in the perfect mixture are given by the equations[1]

$$\mu_A = \left(\frac{\partial A}{\partial N}\right)_{T,V,M} = \mu_A^\circ + kT \ln x, \qquad (11,40)$$

and

$$\mu_B = \left(\frac{\partial A}{\partial M}\right)_{T,V,N} = \mu_B^\circ + kT \ln (1 - x) \qquad (11.41)$$

which are the same as (11.2) and 11.3), so that

$$P_A = xP_A^0 \quad \text{and} \quad P_B = (1 - x)P_B^0 \qquad (11.42)$$

which is Raoult's law.

[1] When proceeding from (11.39) to (11.40) and (11.41) the reader should take account of the fact that x is a function of both N and M.

We have now deduced, using the methods of statistical mechanics, all relations associated with the perfect mixture. It is timely therefore to remind the reader of the conditions under which the above equations were deduced:

(i) that the molecules A and B are non-polar and of approximately the same size and shape, and

(ii) that the forces of attraction between unlike molecules are the arithmetic mean of those between like molecules A and like molecules B, so that the parameter w' is zero. (We take this opportunity of emphasising that condition (ii) does *not* imply that the parameters ϵ_{AA} and ϵ_{BB} are the same. No such assumption was implied in the arguments of either the present section or the last.)

It is probable that mixtures of molecules which differ only by isotopic substitution are the only ones which comply precisely with these requirements, and it is not clear from the viewpoint of statistical theory why systems with greater differences in molecular properties yield perfect mixtures within the limits of experimental error. It is possible that small differences in molecular size and small differences in inter-molecular forces produce opposite deviations from Raoult's law, and that opposing effects cancel.

11.5   The Statistical Mechanics of Mixtures Characterised by the Condition $w' \neq 0$

We now suppose that all the assumptions and conditions made in Section 11.2 remain, but that the parameter w' introduced in equation (11.29) is other than zero. The first problem is that of determining zX, the number of unlike nearest-neighbour pairs in the mixture. It is as well to accept from the start that no method capable of producing an *exact* expression for zX has yet been described, but by making an assumption which is obviously not quite true, an approximate value for zX is easily obtained.

We have N molecules of A and M molecules of B to arrange on $N + M$ equivalent sites. We suppose that each molecule is surrounded by z nearest neighbours so that there are $\frac{1}{2}z(N + M)$ pairs of adjacent sites in all. Consider two adjacent sites 1 and 2 on each of which a molecule is located. The probability that the molecule located on site 1 is of species A is $N/(N + M)$. We now consider the probability that the molecule located on site 2 is of species B. *If we suppose that a molecule B 'exercises no preference' between being located next to another of like kind or next to another of unlike kind* the probability that a molecule B settles on site 2 is $M/(N + M)$, so that the probability that an A molecule is located on site 1 and that, at the same time, a B molecule is located on site 2 is, because these separate events are supposedly independent, the product of the two terms

$$\frac{N}{N + M} \quad \text{and} \quad \frac{M}{N + M}, \quad \text{i.e.}$$

$$\frac{NM}{(N + M)^2}.$$

The probability that a B molecule is located on site 1 and an A molecule on site 2 is the same, but this situation is of course independent of the first (because the two situations $A_{(1)}B_{(2)}$ and $A_{(2)}B_{(1)}$ are mutually exclusive), so that the chance that a particular pair of adjacent sites is occupied by an unlike pair of molecules irrespective whether the arrangement is $A_{(1)}B_{(2)}$ or $A_{(2)}B_{(1)}$ is

$$\frac{2NM}{(N + M)^2}.$$

There are $\frac{1}{2}z(N + M)$ pairs of adjacent sites in all, so that the probable number of unlike pairs in the mixture is

$$\tfrac{1}{2}z(N + M) \cdot \frac{2NM}{(N + M)^2}.$$

Thus,

$$zX = z\,\frac{NM}{N + M}. \tag{11.43}$$

When considering the perfect mixture ($w' = 0$), we correctly assumed that

$$\frac{W_{(c)}}{W_{(c)A} \cdot W_{(c)B}} = \frac{(N + M)!}{N!\,M!}. \tag{11.44}$$

We now assume that this relation holds for the case in which w' is other than zero, and substituting (11.43) and (11.44) into (11.34) find that

$$\Delta A_m = -kT \ln \frac{(N + M)!}{N!\,M!} + z\,\frac{NM}{N + M}\,w', \tag{11.45}$$

so that

$$\Delta A_m = NkT \ln x + MkT \ln (1 - x) + z\,\frac{NM}{N + M}\,w'. \tag{11.46}$$

We hasten to remark that neither the assumption upon which the argument resulting in (11.43) was based (that assumption shown in italics), nor the replacement of

$$\frac{W_{(c)}}{W_{(c)A} \cdot W_{(c)B}} \quad \text{by} \quad \frac{(N + M)!}{N!\,M!}$$

is really valid if w' is other than zero. It is apparent from elementary considerations that if w' is positive, the value of zX given by (11.43) will be somewhat too high, as will be the last term in (11.46), and if w' is negative, the value of zX given by (11.43) will be somewhat too low, but that, since w' is negative, the value of the last term in (11.46) will again be too high. Similar considerations show that for all values of w' other than zero, the distribution in the mixture will be less random than that implied by the expression $(N + M)!/N!M!$, so that the term $-kT \ln(N + M)!/N!M!$ will be somewhat too small. We now know that neither error is likely to be large, (although the first is thought to be more important than the second), but we observe that in all cases the second goes some way towards cancelling the first.

Making the same assumption as that made in the last section, that

$$A_A \simeq N\mu_A^\circ \quad \text{and} \quad A_B \simeq M\mu_B^\circ,$$

it follows from (11.46) that

$$A_{\text{mixture}} = N\mu_A^\circ + M\mu_B^\circ + NkT \ln x + MkT \ln(1 - x)$$

$$+ z \frac{NM}{N + M} w', \tag{11.47}$$

so that the chemical potentials of A and B in the mixture are given by the equations

$$\mu_A = \left(\frac{\partial A}{\partial N}\right)_{T,V,M} = \mu_A^\circ + kT \ln x + zw'(1 - x)^2 \tag{11.48}$$

$$\mu_B = \left(\frac{\partial A}{\partial M}\right)_{T,V,N} = \mu_B^\circ + kT \ln(1 - x) + zwx^2. \tag{11.49}$$

Comparison of (11.48) with (11.8) and (11.49) with (11.9) shows that

$$\ln \gamma_A = \frac{zw'}{kT}(1 - x)^2$$

and $\hspace{11cm}$ (11.50)

$$\ln \gamma_B = \frac{zw'}{kT} x^2.$$

We now, for convenience introduce a parameter w defined by the equation

$$w = Lzw', \tag{11.51}$$

where L is, as always, the Avogadro constant, and write

$$\ln \gamma_A = \frac{w}{RT}(1 - x)^2, \tag{11.52}$$

and

$$\ln \gamma_B = \frac{w}{RT} x^2. \tag{11.53}$$

These equations are the same as the empirical relations shown in (11.18) if we write

$$b = \frac{w}{RT}.$$

It follows also, that since for any mixture

$$\frac{P_A}{P_A^0} = \gamma_A x \quad \text{and} \quad \frac{P_B}{P_B^0} = \gamma_B (1 - x),$$

the statistical–mechanical expressions for the partial vapour pressures are

$$P_A = P_A^0 x \exp\left[\frac{w}{RT}(1 - x)^2\right], \tag{11.54}$$

and

$$P_B = P_B^0 (1 - x) \exp\left[\frac{w}{RT} x^2\right]. \tag{11.55}$$

We observe that if w is positive, the mixture shows positive deviations from Raoult's law, and negative deviations if w is negative.

Despite the many assumptions made we have, perhaps surprisingly, arrived at theoretical expressions for activity coefficients and partial pressures which appear to accord precisely with those of the regular mixture. A more searching test of our theory requires, however, an expression for the molar energy of mixing.

Starting with equation (11.46), substituting xL for N, $(1 - x)L$ for M, and making use of the parameter w defined by (11.51) we find that

$$\Delta A_m^m = xRT \ln x + (1 - x)RT \ln(1 - x) + x(1 - x)w, \tag{11.56}$$

from which

$$\Delta E_m^m = \Delta A_m^m - T\left(\frac{\partial(\Delta A_m^m)}{\partial T}\right)_V = x(1 - x)\left[w - T\frac{dw}{dT}\right]. \tag{11.57}$$

In the last step we have assumed that the parameter w is dependent on temperature, there being nothing in the definition of w' (from which w derives) to indicate otherwise. In fact such experimental evidence as we have (see Section 11.7) indicates that if w is positive, dw/dT is negative, so that ΔE_m^m is positive. We saw earlier that if w is positive the mixture shows positive deviations from Raoult's law. We now deduce that positive deviations result from endothermic mixing processes. The position regarding the case in which w is negative is less certain for the very good reason that we know of no mixture which conforms closely to the conditions imposed in Section 11.2, and which shows negative deviations from Raoult's law. Such evidence as we have suggests that negative deviations are associated with exothermic mixing processes.

11.6   Incomplete Miscibility

It will now be demonstrated that the relations established in the last section are entirely consistent with the experimental evidence relating to the phenomenon of incomplete miscibility. We require only the expression for ΔA_m^m given in (11.56) which is written in the form

$$\frac{\Delta A_m^m}{RT} = x \ln x + (1 - x) \ln (1 - x) + x(1 - x) \frac{w}{RT}. \qquad (11.58)$$

The first two terms on the right-hand side are negative, and the third positive or negative depending on w.

Curves showing $\Delta A_m^m/RT$ against x for various values of w/RT are shown in Figure 11.2. It is evident by inspection that whereas in curves (1) $(w/RT = 0)$, (2) $(w/RT = 1.0)$, and (5) $(w/RT = -1.0)$, and indeed, in curves for all *negative* values of w/RT, there exists only one value of x $(x = 0.5)$ at which $\partial(\Delta A_m^m)/\partial x$ is zero, three such values exist in the case of curve (3) $(w/RT = 2.5)$. It is evident also that a homogeneous mixture the composition of which lies between points J and K on curve (3) may decrease the value of its Helmholtz function and so increase its stability, by separating into a pair of immiscible conjugate solutions of composition J and K respectively, (the amount of each phase being determined by the overall composition of the mixture). It is thus apparent that the phenomenon of partial miscibility is associated with curves of type (3), and therefore with positive values of w/RT greater than some critical value. It is also apparent from the geometry of curve (3) that there are two values of x at which $\partial^2(\Delta A_m^m)/\partial x^2 = 0$, and that these coincide in the curve for the critical value of w/RT. Such a situation is given by the condition $\partial^3(\Delta A_m^m)/\partial x^3 = 0$. It follows from (11.58) that

$$\frac{\partial^2(\Delta A_m^m)}{\partial x^2} = \frac{RT}{x(1 - x)} - 2w,$$

and that

$$\frac{\partial^3(\Delta A_m^m)}{\partial x^3} = \frac{2x - 1}{x^2(1 - x)^2}.$$

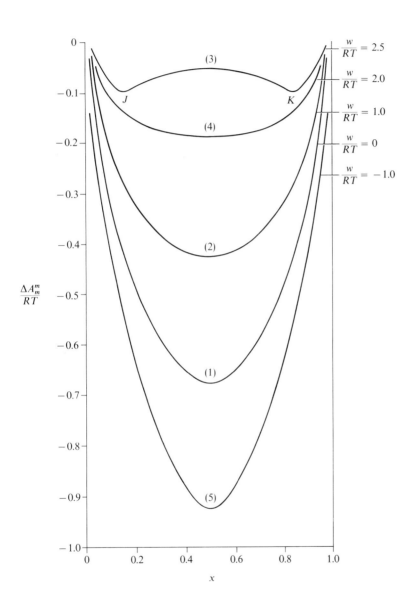

Figure 11.2 $\dfrac{\Delta A_m^m}{RT}$ for various values of $\dfrac{w}{RT}$.

Equating both to zero, we find that the critical value for w/RT is given by the equation

$$\frac{w}{RT} = 2. \tag{11.59}$$

The curve for $\Delta A_m^m/RT$ for this value of w/RT is shown as curve (4).

Our equations suggest therefore that if the parameter w is such that for a mixture at temperature T, w is less than $2RT$, we would expect complete miscibility, (and therefore that mixtures showing negative values of w should be miscible over the entire composition range,) but that if w is greater than $2RT$, incomplete miscibility would be exhibited. We must now demonstrate how it comes about that a mixture which is completely miscible at and above one temperature may exhibit limited miscibility at lower temperatures.

We first observe that

$$\frac{d(w/RT)}{dT} = \frac{1}{RT^2}\left[T\frac{dw}{dT} - w \right]. \tag{11.60}$$

We have already seen that limited miscibility is associated only with positive values of w, and in the last section stated that it appears that positive values of w are associated with negative values for dw/dT. It follows from (11.60) that if w is positive $d(w/RT)/dT$ is negative, so that decrease in T leads to an increase in w/RT. If the decrease in temperature is such that w/RT changes from a value less than two to a value greater than two, limited miscibility will be expected at the lower temperature.

Some pairs of liquids are known which are completely miscible above one temperature (called the upper consolute temperature), and below a lower temperature (called the lower consolute temperature), but exhibit limited miscibility in the intermediate range. An example of such a pair is nicotine and water, the two temperatures being 481 K and 334 K. Guggenheim[1] has shown that such behaviour is possible in the case of mixtures obeying the equations of Section 11.5, if the dependence of w on T has the form

$$\frac{w}{R} = 2T + \frac{a^2 - (T - b)^2}{c}$$

where a, b and c are positive constants and a is less than b, the two consolute temperatures having the values $b + a$ and $b - a$. Other pairs of liquids, such as tri-ethylamine and water are completely miscible below a particular temperature, (291 K in the example given,) but show limited miscibility at higher temperatures. Presumably a mixture could exhibit this kind of behaviour if the dependence of w on T has the form shown above, but a and

[1] Guggenheim, *Faraday Soc. Discussions* **15**,, 271 (1953).

b have such values that the higher consolute temperature $b + a$ is above the temperature at which the mixture can exist as a liquid.

11.7 Comparison with Experiment and More Sophisticated Treatments

The most searching test of the theoretical expressions produced in Section 11.5 is obtained:

(i) by choosing for different temperatures, values of w which provide the best fit between the theoretical values of ΔA_m^m given by equation (11.56) and the experimental values of ΔG_m^m obtained via equation (11.13) from measurements of partial vapour pressures, so yielding values for dw/dT, and

(ii) using these values and those of w to give a theoretical value for ΔE_m^m via equation (11.57) for comparison with experimental values of heats of mixing.

In fairness to the theory, we must choose mixtures only of non-polar molecules, and those of approximately the same size. These conditions are satisfied reasonably well by mixtures of carbon tetrachloride and cyclo-hexane, (the molar volumes of which are $97 \, cm^3$ and $109 \, cm^3$ respectively), and by mixtures of carbon tetrachloride and benzene, (the molar volume of benzene is $89 \, cm^3$.) The first mixture was investigated by Adcock and McGlashan[1] using their own experimental values for heats of mixing, and values for ΔG_m^m obtained by Scatchard, Wood and Mochel.[2] It was found that both sets of data correspond to the relations

$$w/J \, mol^{-1} = 1176 + 1.96 \, T \ln T/K - 14.18T,$$

and

$$\left[w - T \frac{dw}{dT} \right] \Big/ J \, mol^{-1} = 1176 - 19.6T.$$

These expressions lead to the values

$$w = 276 \, J \, mol^{-1} \text{ at } 300 \, K,$$

and

$$\frac{dw}{dT} = -1.04 \, JK^{-1} \, mol^{-1}$$

at the same temperature. Details of an earlier analysis of the system carbon

[1] Adcock and McGlashan, *Proc. Roy. Soc.* **A226**, 266 (1954).

[2] Scatchard, Wood and Mochel, *J. Amer. Chem. Soc.* **61**, 3206 (1939).

tetrachloride/benzene are given by Guggenheim.[1] A good fit between the values of ΔG_m^m and ΔA_m^m is obtained if values for w are chosen which change regularly from 325 J mol^{-1} at $30°C$ to 300 J mol^{-1} at $70°C$, so yielding the value $dw/dT = -0.625$ JK^{-1} mol^{-1}. These figures yield a calculated value for ΔE_m^m at $25°C$ for the equimolar mixture of 130 J mol^{-1}. An experimental value for the heat of mixing at this temperature is 109 JK^{-1} mol^{-1}.

Notwithstanding the apparent success implied by these results, it must be clear that some of the assumptions made in Section 11.5 might well be improved, and the attempts to do so are legion. Most have centred around the problem of determining zX, the most probable number of unlike pairs, in order to obtain a better value than that given by the equation

$$zX = z\frac{NM}{N + M},$$
(11.43)

the derivation of which has already been criticised. This equation is equivalent to the expression

$$X^2 = (N - X)(M - X),$$
(11.61)

as may easily be verified; and the first major refinement, due originally to Guggenheim, and known as the quasi-chemical treatment, was to replace (11.61) by the equation

$$X^2 = (N - X)(M - X)\exp\left[-\frac{2w}{zRT}\right].$$
(11.62)

Further refinements were attempted by taking into account interference between pairs of sites, and by including in the expressions, terms relating to the attractive forces between 'next-to-nearest neighbours.' An exhaustive account of these refinements is given in Guggenheim's monograph[1] to which the interested reader is referred. It is perhaps fair to remark that few of the experimental data available are sufficiently accurate to provide a fitting test of some of the refinements made.

11.8   Relaxation of the Restrictions Imposed on the Lattice Model

In earlier sections we have, by analysing a model subject to the restrictions imposed in Section 11.2, provided a reasonably-satisfactory statistical–mechanical foundation to the chemistry of perfect mixtures, and to that of

[1] Guggenheim, *Mixtures* (Oxford, 1952).

regular mixtures, (those for which the activity coefficients accord to equations (11.18.) It is therefore reasonable to suppose that the reason why most imperfect mixtures are governed by equations (11.16) rather than (11.18) stems from the departure of these mixtures from one or other of the restrictions imposed in Section 11.2.

Of these, the most important are:

(i) the assumption that the molecules of both species are of approximately the same size and shape, and

(ii) the assumption that the molecules of both species are non-polar.

We shall now mention, very briefly indeed, some of the attempts which have been made to relax these restrictions.

Molecules of unequal size

The principal difficulty in applying a quasi-crystalline lattice model to mixtures of molecules of unequal size lies in determining the number of ways in which N molecules of one species and M of the other can be arranged. This problem was first tackled by Fowler and Rushbrooke[1] who studied the case in which it was supposed that one molecule (A) is spherical, and that the other (B) consists of two spheres each of the same size as the first but rigidly joined to form a dumb-bell. This is, of course, the simplest possible case, because we might suppose that the replacement of two molecules of A in the lattice by one molecule of B might be made without strain, and secondly, we may suppose that in a mixture of N molecules of A and M molecules of B, we have $N + 2M$ lattice sites, each molecule of A occupying one, and each molecule of B occupying two neighbouring sites. Even for this relatively simple case, the formulae for the number of ways in which the molecules may be arranged, and for the thermodynamic quantities of mixing are very complicated. The most important result of this treatment was to show that such mixtures (for the value $w' = 0$) would exhibit small *negative* deviations from Raoult's law.

An extreme case of molecules of unequal size is of course offered by solutions of high polymers, and much of our understanding of the behaviour of such solutions has been obtained from studies made on quasi-crystalline lattice models. This is a field of chemistry which has received so much attention in the last thirty years, that no worthwhile description can be encompassed in a small space. For study of this topic, the reader is referred

[1] Fowler and Rushbrooke, *Trans. Faraday Soc.* **33**, 1272 (1937).

first to a review by Patterson[1], and then to the monographs by Flory,[2] Tompa[3] and Morawetz.[4]

Polar molecules

Some attempts have been made, by Tompa,[5] and by Barker and his associates,[6] to extend the quasi-crystalline lattice treatment to cover mixtures characterised by strong highly-localised interactions such as hydrogen bonds. These attempts consist essentially (i) of constructing expressions for the quantities of mixing in terms of the various possible interaction energies concerned, (ii) choosing values for these so that the theoretical values for the thermodynamic quantities of mixing at mole fraction $x = 0.5$ are the same as the experimental values, and (iii) of using these values to obtain theoretical curves for these quantities over the entire composition range. Many of the experimental curves for these quantities for mixtures in which strong interactions may be expected are highly asymmetric. Barker's theoretical curves reproduce this asymmetry very well, although precise correlation is not obtained. Nor indeed, considering the complexity of the problem, is it to be expected.

11.9   Dilute Solutions

We end this chapter by considering briefly, liquid mixtures in which component B is in very great excess. In these circumstances it is usual to refer to component B as the *solvent*, component A as the *solute*, and the mixture as a solution of A in B. By definition therefore, when speaking of a *dilute* solution of A in B, we mean a system in which x, the mole fraction of solute is very close to zero, so that $1 - x$, the mole fraction of solvent is very close to unity, and N, the number of molecules of A is *very* much smaller than M, the number of molecules of B.

The most interesting property of dilute solutions as far as we are here concerned, is that no matter how imperfect is the mixture when viewed over the entire composition range, the partial pressure of B over the range in which x is very small is given by the equation

$$P_B = (1 - x)P_B^0,$$ (11.63)

[1] Patterson, *Rubber and Science Technology* **40**, 1 (1967.

[2] Flory, *Principles of Polymer Chemistry* (Cornell University Press, 1953).

[3] Tompa, *Polymer Solutions* (Butterworths, 1954).

[4] Morawetz, *Macromolecules in Solution* (Interscience, 1965).

[5] Tompa, *J. Chem. Phys.* **21**, 250 (1953), *Disc. Faraday Soc.* **15**, 259 (1953).

[6] Barker *et al.*, *J. Chem. Phys.* **20**, 1526 (1952); **21**, 1391 (1953); **22**, 375 (1954); *Disc. Faraday Soc.* **15**, 142 (1953).

(so that the solvent obeys Raoult's law,) but that that of A is given by the equation

$$P_A = Cx, \qquad (11.64)$$

where (compare equation 11.1) the proportionality constant C is *not* equal to P_A^o. Equation (11.64) is called Henry's law, and the proportionality constant C, the Henry constant, and dilute solutions in which the solvent obeys Raoult's law, and the solute Henry's law are called *ideal*.

Expressions for the chemical potentials of solvent and solute are easily obtained. Considering the equilibrium between both components in the liquid phase and those in the vapour, we have

$$\mu_{B(liq)} = \mu_{B(g)} = \mu_B^\dagger + kT \ln \frac{P_B}{P^\dagger}$$

$$= \mu_B^\dagger + kT \ln \frac{P_B^0}{P^\dagger} + kT \ln (1 - x)$$

$$= \mu_B^o + kT \ln (1 - x), \qquad (11.65)$$

which is of course, the same as (11.3), but

$$\mu_{A(liq)} = \mu_{A(g)} = \mu_A^\dagger + kT \ln \frac{P_A}{P^\dagger}$$

$$= \mu_A^\dagger + kT \ln \frac{C}{P^\dagger} + kT \ln x$$

$$= \mu_A^\ominus + kT \ln x, \qquad (11.66)$$

where we define the quantity $\mu_A^\ominus$ by the equation

$$\mu_A^\ominus = \mu_A^\dagger + kT \ln \frac{C}{P^\dagger}. \qquad (11.67)$$

The important point is that $\mu_A^\ominus$ is *not* equal to μ_A^o (the chemical potential of A as a pure liquid at the same temperature) because

$$\mu_A^o = \mu_A^\dagger + kT \ln \frac{P_A^0}{P^\dagger}$$

and C is not equal to P_A^0.

It is of obvious interest to seek a statistical–mechanical foundation for equation (11.66), that is, an explanation of the fact that Henry's law is invariably obeyed by very dilute solutions of non-electrolytes.[1] An explanation follows immediately from the approach described in Section 11.5.

We start with equation (11.34) and suppose that w' is other than zero. We again assume that we may replace the quotient

$$\frac{W_{(c)}}{W_{(c)A} \cdot W_{(c)B}} \quad \text{by} \quad \frac{(N + M)!}{N! \, M!} \, ,$$

and so obtain the equation

$$\Delta A_m = NkT \ln x + MkT \ln (1 - x) + zXw', \tag{11.68}$$

where z, X and w' have the same significance as before.

To proceed we require a value for zX, but now there is no problem whatsoever, because, since N is very much smaller than M, it is almost certain that each molecule of solute will be surrounded completely by molecules of solvent, so that every molecule of solute will participate in z unlike pairs, and zX will be the same as zN.

Equation (11.68) therefore becomes

$$\Delta A_m = NkT \ln x + MkT \ln (1 - x) + zNw'. \tag{11.69}$$

The Helmholtz function for the solution is given by the equation

$$A_{\text{solution}} = N\mu_A^\circ + NkT \ln x + M\mu_B^\circ + MkT \ln (1 - x) + zNw', \tag{11.70}$$

from which it follows that

$$\mu_B = \left(\frac{\partial A}{\partial M}\right)_{T,V,N} = \mu_B^\circ + kT \ln (1 - x), \tag{11.71}$$

which is the same as (11.65), but

$$\mu_A = \left(\frac{\partial A}{\partial N}\right)_{T,V,M} = \mu_A^\circ + zw' + kT \ln x. \tag{11.72}$$

Equation (11.72) becomes identical with (11.66) if we *define* $\mu_A^\ominus$ by the equation

$$\mu_A^\ominus = \mu_A^\circ + zw'. \tag{11.73}$$

It follows from (11.67) that the Henry constant C is given by the equation

$$kT \ln \frac{C}{P^\dagger} = \mu_A^\circ + zw' - \mu_A^\dagger. \tag{11.74}$$

---

[1] We have to insert the words 'of non-electrolytes' because dilute solutions of electrolytes do *not* obey Henry's law.

EXERCISES

11.1 If in a binary mixture it is found experimentally that

$$\ln \gamma_A = b(1 - x)^2 + c(1 - x)^3$$

and

$$\ln \gamma_B = b'x^2 + c'x^3,$$

where x is the mole fraction of component A, show that it follows from the Gibbs–Duhem equation (11.14) that

$$b' = b + \tfrac{3}{2}c \quad \text{and} \quad c' = c.$$

Allot possible values to b and c, e.g. $b = 0.2$, $c = 0.02$ and plot $\gamma_A x$ (i.e. P_A/P_A°) and $\gamma_B(1 - x)$ (i.e. P_B/P_B°) against x from $x = 0$ to $x = 1$.

11.2 A particular mixture of liquids of *unequal* molecular volumes shows a positive heat of mixing, but the partial vapour pressures of its components appear to be in accord with Raoult's law. Show how these facts may be reconciled.

11.3 Show using the equations of Section 11.9 that the Henry law constant C is greater than P_A° if w' is positive, and less than P_A° if w' is negative.

CHAPTER 12

The Grand Canonical Ensemble

12.1 Introduction

The application of statistical mechanics to many problems is facilitated by the use of the relations of the grand canonical ensemble. This is an isolated collection of very many identical macroscopic systems, each of which is of fixed volume, but supposedly *open*, so that both energy and material can pass from one system to another. No restrictions of any kind are imposed on the systems of the ensemble: each may consist of more than one phase, and each may be an assembly of interacting particles.

The relations of the grand canonical ensemble are no more powerful than those of the canonical ensemble, in the sense that no problem which is not amenable to the methods of the latter is amenable to the methods of the first. Thus, the grand canonical partition function Z_G can be *evaluated* only in the same circumstances as can the canonical partition function Q, i.e. for systems of independent particles. But the relations to be developed in the present chapter, complicated as they may appear at first sight, prove, in many cases to be easier to apply than those of the canonical ensemble, and very much easier to apply than those of the micro-canonical ensemble.

One reason for this is that the relations of the grand canonical ensemble, being framed for *open* systems, are directly applicable to studies of (necessarily) *open phases* within a multiphase system. This point will be illustrated in the next chapter where we re-examine, using the methods of the grand canonical ensemble, the topic of the adsorption of gases, previously examined, using the methods of the canonical ensemble, in Chapter 10.

12.2 The Construction of the Grand Canonical Partition Function

We start by considering an ensemble of systems containing two components A and B. This renders our equations slightly more complicated than they would be had we chosen an ensemble of systems containing a single component, but has the advantage that the extension of all relations to single- or multi-component systems is then obvious.

Consider therefore, an isolated ensemble of $\mathcal{N}$ identical macroscopic systems, each of constant volume V, so that $\mathscr{V}$, the volume of the ensemble is $\mathcal{N}V$, and suppose that each system contains molecules of two kinds, and is permitted to exchange both energy and molecules with its neighbours. Denote the energy of the ensemble and the numbers of molecules of A and B therein by the symbols $\mathscr{E}$, $\mathbb{N}$ and $\mathbb{M}$, and the average energy and average number of molecules of each kind per system by the symbols $\bar{E}$, $\bar{N}$ and $\bar{M}$. We then have the relations

$$\mathscr{E} = \mathcal{N}\bar{E}, \tag{12.1}$$

$$\mathbb{N} = \mathcal{N}\bar{N} \tag{12.2}$$

and

$$\mathbb{M} = \mathcal{N}\bar{M}. \tag{12.3}$$

We shall ultimately require an expression for Ω_{total}, the number of quantum-mechanical states accessible to the ensemble. This is rather more complicated to construct than it was in the case of the canonical ensemble, because there, the material content of each system was fixed (and the same in each) and only the energy of each was supposedly capable of change. In the present case each system can acquire different energies *and* different amounts of each component. We shall frame our expression for Ω_{total} in terms of a quantity $\mathcal{N}_{N,M,j}$, the number of systems which contains N molecules of A, M molecules of B and which are in *the jth quantum-mechanical state for these numbers of molecules.* It is the meaning of the last phrase with which we are first concerned.

The permissible energy levels accessible to a *closed* thermostatted system are determined by its volume *and its material content*, so that the set of energies $E_{N,M,V}$ accessible to a thermostatted system of volume V containing N molecules of A and M molecules of B differs from the set of energies $E_{N',M',V}$ accessible to a system of the same volume but containing different numbers of molecules, i.e. N' of A and M' of B. Each system characterised by N molecules of A and M molecules of B and *one* of its permissible energies $E_{N,M,V}$ may, of course, exist in any one of W quantum-mechanical states, where W is determined solely by N, M, V and the chosen value for $E_{N,M,V}$.

The symbol $\mathcal{N}_{N,M,j}$ denotes the number of systems in the ensemble in the jth of these quantum-mechanical states.

The total number of systems, the total energy of the ensemble and the total number of molecules of each kind may now be expressed in terms of $\mathcal{N}_{N,M,j}$, the appropriate equations being

$$\mathcal{N} = \sum_N \sum_M \sum_j \mathcal{N}_{N,M,j}, \tag{12.4}$$

$$\mathcal{E} = \sum_N \sum_M \sum_j \mathcal{N}_{N,M,j} E_{N,M,j} \tag{12.5}$$

where $E_{N,M,j}$ is the energy characterising the jth state,

$$\mathbb{N} = \sum_N \sum_M \sum_j \mathcal{N}_{N,M,j} N \tag{12.6}$$

and

$$\mathbb{M} = \sum_N \sum_M \sum_j \mathcal{N}_{N,M,j} M. \tag{12.7}$$

In each case summation is carried out for each j for each of all possible values of N and M.

We point out immediately the essential difference between the term $\mathcal{N}_{N,M,j}$ as here used, and the term $\mathcal{N}_i$ used in Chapter 8. In that chapter we chose to denote by the symbol $\mathcal{N}_i$ the number of systems characterised by energy E_i, so that $\mathcal{N}_i$ includes all systems in the $W(E_i)$ quantum states characterised by energy E_i, and in such equations as (8.9) and (8.10) we summed over all energy *levels*. In the present chapter we choose to sum over all quantum-mechanical *states*.

Since $\mathcal{N}$, $\mathcal{E}$, $\mathbb{N}$ and $\mathbb{M}$ are fixed, all possible distributions of systems over states are governed by the expressions

$$\sum_{N,M,j} d\mathcal{N}_{N,M,j} = 0, \tag{12.8}$$

$$\sum_{N,M,j} E_{N,M,j} d\mathcal{N}_{N,M,j} = 0, \tag{12.9}$$

$$\sum_{N,M,j} N\, d\mathcal{N}_{N,M,j} = 0 \tag{12.10}$$

and

$$\sum_{N,M,j} M\, d\mathcal{N}_{N,M,j} = 0, \tag{12.11}$$

where for notational brevity we denote the triple sum by the symbol $\sum_{N,M,j}$.

Since each system is macroscopic, and hence in principle may be 'labelled' and so distinguished from its neighbours, any one distribution of systems over states may be reached in Ω ways where

$$\Omega = \frac{\mathcal{N}!}{\prod_{N,M,j} \mathcal{N}_{N,M,j}!} \tag{12.12}$$

and where $\prod_{N,M,j}$ denotes the running product of each $\mathcal{N}_{N,M,j}!$. There are, of course, very many possible distributions (i.e. many different sets of distribution numbers $\mathcal{N}_{N,M,j}$) each described by an equation similar to (12.12), so that the total number of quantum-mechanical states accessible to the ensemble is given by the expression

$$\Omega_{\text{total}} = \frac{\mathcal{N}!}{\prod_{N,M,j} \mathcal{N}_{N,M,j}!} + \frac{\mathcal{N}!}{\prod_{N,M,j} \mathcal{N}'_{N,M,j}} + \cdots .$$

We realise however that, so long as $\mathcal{N}$ is large, $\ln \Omega_{\text{total}}$ is indistinguishable from the logarithm of the Ω value for the most probable distribution, and that the most probable values for all $\mathcal{N}_{N,M,j}$ are those satisfying the equation

$$d \ln \Omega = \sum_{\mathcal{N}_{N,M,j}} \frac{\partial \ln \Omega}{\partial \mathcal{N}_{N,M,j}} d\mathcal{N}_{N,M,j} = 0,$$

subject to the restrictions imposed by equations (12.8) to (12.11). It follows from (12.12) that

$$\ln \Omega = \mathcal{N} \ln \mathcal{N} - \sum_{\mathcal{N}_{N,M,j}} \left[\mathcal{N}_{N,M,j} \ln \mathcal{N}_{N,M,j} - \mathcal{N}_{N,M,j} \right] \tag{12.13}$$

so that

$$\frac{\partial \ln \Omega}{\partial \mathcal{N}_{N,M,j}} = -\ln \mathcal{N}_{N,M,j}$$

and

$$d \ln \Omega = - \sum_{\mathcal{N}_{N,M,j}} \ln \mathcal{N}_{N,M,j} \, d\mathcal{N}_{N,M,j} = 0. \tag{12.14}$$

We combine (12.14) with conditions (12.8) to (12.11) by multiplying (12.8) by α, (12.9) by $-\beta$, (12.10) by $\ln \lambda_A$ and (12.11) by $\ln \lambda_B$, the last two multipliers being so chosen for future convenience, and adding to (12.14) obtain the single condition

$$\sum_{\mathcal{N}_{N,M,j}} \left[-\ln \mathcal{N}_{N,M,j} + \alpha - \beta E_{N,M,j} + N \ln \lambda_A + M \ln \lambda_B \right] d\mathcal{N}_{N,M,j} = 0. \tag{12.15}$$

The quantity in brackets must be zero for all $\mathcal{N}_{N,M,j}$, so that the most probable distribution numbers are those given by the set of equations

$$\ln \mathcal{N}_{N,M,j} = \alpha - \beta E_{N,M,j} + N \ln \lambda_A + M \ln \lambda_B;$$

i.e.

$$\mathcal{N}_{N,M,j} = \exp \alpha \cdot \lambda_A^N \cdot \lambda_B^M \cdot \exp(-\beta E_{N,M,j}). \qquad (12.16)$$

We may immediately eliminate $\exp \alpha$, since from (12.4)

$$\mathcal{N} = \exp \alpha \cdot \sum_{N,M,j} [\lambda_A^N \lambda_B^M \exp(-\beta E_{N,M,j})],$$

so that

$$\frac{\mathcal{N}_{N,M,j}}{\mathcal{N}} = \frac{\lambda_A^N \lambda_B^M \exp(-\beta E_{N,M,j})}{\sum_{N,M,j} [\lambda_A^N \lambda_B^M \exp(-\beta E_{N,M,j})]}$$

$$= \frac{\lambda_A^N \lambda_B^M \exp(-\beta E_{N,M,j})}{Z_G}, \qquad (12.17)$$

where

$$Z_G = \sum_{N,M,j} [\lambda_A^N \lambda_B^M \exp(-\beta E_{N,M,j})]. \qquad (12.18)$$

The quantity Z_G is called the grand canonical partition function.[1] It is as well to pause for a moment to consider the physical significance of equation (12.17). This equation gives the number of systems which contain precisely N molecules of one kind and M of the other and which are in quantum state j. Had we chosen (as we did when constructing the particle and canonical partition functions) to sum over energy *levels* E_i (rather than over quantum *states*) we should have denoted by the symbol $\mathcal{N}_{N,M,E_i}$ the number of systems containing N molecules of one kind and M of the other in *all* of the $W(N, M, E_i)$ states characterised by N, M, E_i. The expression for the partition function would then have been written

$$Z_G = \sum_{N,M,E_i} [W(N, M, E_i)\lambda_A^N \lambda_B^M \exp(-\beta E_i)], \qquad (12.19)$$

(it should be clear that the numerical values of the sums in (12.18) and (12.19) are the same,) and the corresponding expression to (12.17) would be

$$\frac{\mathcal{N}_{N,M,E_i}}{\mathcal{N}} = \frac{W(N, M, E_i)\lambda_A^N \lambda_B^M \exp(-\beta E_i)}{Z_G}.$$

12.3 Expressions for Ensemble Averages

We now obtain expressions for the ensemble averages, $\overline{N}$, $\overline{M}$, $\overline{E}$, $\overline{S}$ and $\overline{A}$ in terms of Z_G. In so doing we shall identify the multiplier β, and in the

---

[1] The grand canonical partition function is often given the symbol Ξ (pronounced 'zy'). This is tedious to express in writing, so we here use the symbol Z_G the suffix G being appended because in some texts the symbol Z is used for the canonical partition function Q.

next section, when obtaining expressions for the chemical potentials of the components A and B, find that we have identified the multipliers $\ln \lambda_A$ and $\ln \lambda_B$.

It is convenient first to differentiate Z_G with respect to λ_A, regarding β, V, and λ_B as constant. (Since the possible energies are determined by N, M and V, constancy of V implies constancy of the energies $E_{N,M,j}$.) Thus

$$\left(\frac{\partial Z_G}{\partial \lambda_A}\right)_{\beta,V,\lambda_B} = \sum_{N,M,j} N\lambda_A^{N-1}\lambda_B^M \exp\left(-\beta E_{N,M,j}\right). \tag{12.20}$$

Similarly,

$$\left(\frac{\partial Z_G}{\partial \lambda_B}\right)_{\beta,V,\lambda_A} = \sum_{N,M,j} \lambda_A^N M\lambda_B^{M-1} \exp\left(-\beta E_{N,M,j}\right). \tag{12.21}$$

Lastly we differentiate Z_G with respect to β, regarding λ_A, λ_B and V as constant, to obtain

$$\left(\frac{\partial Z_G}{\partial \beta}\right)_{\lambda_A,\lambda_B,V} = -\sum_{N,M,j} \lambda_A^N\lambda_B^M E_{N,M,j} \exp\left(-\beta E_{N,M,j}\right). \tag{12.22}$$

We now find, using (12.6), (12.17) and (12.20) that

$$\bar{N} = \frac{\lambda_A}{Z_G}\left(\frac{\partial Z_G}{\partial \lambda_A}\right)_{\beta,V,\lambda_B} = \left(\frac{\partial \ln Z_G}{\partial \ln \lambda_A}\right)_{\beta,V,\lambda_B}, \tag{12.23}$$

and, using (12.8), (12.17) and (12.21) that

$$\bar{M} = \left(\frac{\partial \ln Z_G}{\partial \ln \lambda_B}\right)_{\beta,V,\lambda_A}, \tag{12.24}$$

We find, using (12.5), (12.17) and (12.22) that

$$\bar{E} = -\frac{1}{Z_G}\left(\frac{\partial Z_G}{\partial \beta}\right)_{\lambda_A,\lambda_B,V} = -\left(\frac{\partial \ln Z_G}{\partial \beta}\right)_{\lambda_A,\lambda_B,V}. \tag{12.25}$$

We now require an expression for $\bar{S}$, so determine first $\mathscr{S}$, the entropy of the ensemble. Obviously,

$$\mathscr{S} = k \ln \Omega_{\max} = k \ln \frac{\mathscr{N}!}{\prod_{N,M,j} \mathscr{N}_{N,M,j}!}.$$

It follows from (12.17) that

$$\frac{\mathscr{S}}{k} = \mathscr{N} \ln \mathscr{N} - \sum_{N,M,j} \mathscr{N}_{N,M,j} \ln \mathscr{N}_{N,M,j}$$

$$= \mathscr{N} \ln \mathscr{N} - \sum_{N,M,j} \mathscr{N}_{N,M,j} \ln \left[\frac{\mathscr{N}\lambda_A^N\lambda_B^M \exp\left(-\beta E_{N,M,j}\right)}{Z_G}\right],$$

so that,

$$\frac{\bar{S}}{k} = \frac{\mathscr{S}}{k\mathscr{N}} = \ln Z_G - \frac{1}{\mathscr{N}} \sum_{N,M,j} \mathscr{N}_{N,M,j} N \ln \lambda_A$$

$$- \frac{1}{\mathscr{N}} \sum_{N,M,j} \mathscr{N}_{N,M,j} M \ln \lambda_B$$

$$+ \frac{\beta}{\mathscr{N}} \sum_{N,M,j} \mathscr{N}_{N,M,j} E_{N,M,j}.$$

Finally, making use of equations (12.6) and (12.2); (12.7) and (12.3) and (12.5) and (12.1), we find that

$$\bar{S} = k \ln Z_G - k\bar{N} \ln \lambda_A - k\bar{M} \ln \lambda_B + \beta k \bar{E}. \tag{12.26}$$

We here digress to identify the Lagrangian multiplier β. The method is similar to that used in Section 2.5, but is given in full because of the slightly greater complexity in the present case. It follows from (12.26) that

$$\left(\frac{\partial \bar{S}}{\partial \bar{E}}\right)_{N,M,V} = k\left(\frac{\partial \ln Z_G}{\partial \bar{E}}\right)_{N,M,V} - k\bar{N}\left(\frac{\partial \ln \lambda_A}{\partial \bar{E}}\right)_{\bar{N},\bar{M},V}$$

$$- k\bar{M}\left(\frac{\partial \ln \lambda_B}{\partial \bar{E}}\right)_{\bar{N},\bar{M},V} + k\bar{E}\left(\frac{\partial \beta}{\partial \bar{E}}\right)_{N,M,V} + \beta k,$$

from which, making use of (12.23) for $\bar{N}$, (12.24) for $\bar{M}$ and (12.25) for $\bar{E}$, we obtain

$$\left(\frac{\partial \bar{S}}{\partial \bar{E}}\right)_{N,M,V} = k\left[\left(\frac{\partial \ln Z_G}{\partial \bar{E}}\right)_{\bar{N},\bar{M},V} - \left(\frac{\partial \ln Z_G}{\partial \ln \lambda_A}\right)_{\beta,\lambda_B,V}\left(\frac{\partial \ln \lambda_A}{\partial \bar{E}}\right)_{\bar{N},\bar{M},V}\right.$$

$$- \left(\frac{\partial \ln Z_G}{\partial \ln \lambda_B}\right)_{\beta,\lambda_A,V}\left(\frac{\partial \ln \lambda_B}{\partial \bar{E}}\right)_{\bar{N},\bar{M},V}$$

$$\left. - \left(\frac{\partial \ln Z_G}{\partial \beta}\right)_{\lambda_A,\lambda_B,V}\left(\frac{\partial \beta}{\partial \bar{E}}\right)_{\bar{N},\bar{M},V}\right] + \beta k. \tag{12.27}$$

Inspection of mathematical theorem 8(ii) on page xvi shows that the quantity in brackets is identically zero,[1] so that

$$\left(\frac{\partial \bar{S}}{\partial \bar{E}}\right)_{\bar{N},\bar{M},V} = \beta k. \tag{12.28}$$

---

[1] Put $C = \ln Z_G$, $X = \ln \lambda_A$, $Y = \ln \lambda_B$, $Z = \beta$, $E = E$, $F = N$, M. Then impose constancy of volume on each term.

Classical thermodynamics shows that for any closed system in equilibrium,

$$\left(\frac{\partial \bar{S}}{\partial \bar{E}}\right)_V = \frac{1}{T},$$

so that

$$\beta = \frac{1}{kT}, \tag{12.29}$$

just as in the cases of the micro-canonical and canonical ensembles.

We may now rewrite equations (12.25) and (12.26) to give

$$\bar{E} = kT^2 \left(\frac{\partial \ln Z_G}{\partial T}\right)_{\lambda_A,\lambda_B,V} \tag{12.30}$$

and

$$\bar{S} = k \ln Z_G - k\bar{N} \ln \lambda_A - k\bar{M} \ln \lambda_B + kT\left(\frac{\partial \ln Z_G}{\partial T}\right)_{\lambda_A,\lambda_B,V} \tag{12.31}$$

from which equations it follows that

$$\bar{A} = \bar{E} - T\bar{S} = -kT \ln Z_G + k\bar{N}T \ln \lambda_A + k\bar{M}T \ln \lambda_B. \tag{12.32}$$

12.4   Identification of λ_A and λ_B and the Criteria for Phase and Chemical Equilibria

It follows from (12.32) that

$$\frac{1}{kT}\left(\frac{\partial \bar{A}}{\partial \bar{N}}\right)_{T,V,\bar{M}} = -\left(\frac{\partial \ln Z_G}{\partial \bar{N}}\right)_{T,V,\bar{M}} + \ln \lambda_A$$
$$+ \bar{N}\left(\frac{\partial \ln \lambda_A}{\partial \bar{N}}\right)_{T,V,\bar{M}} + \bar{M}\left(\frac{\partial \ln \lambda_B}{\partial \bar{N}}\right)_{T,V,\bar{M}},$$

so that, making use of (12.23) and (12.24),

$$\frac{1}{kT}\left(\frac{\partial \bar{A}}{\partial \bar{N}}\right)_{T,V,\bar{M}} = \ln \lambda_A + \left[-\left(\frac{\partial \ln Z_G}{\partial \bar{N}}\right)_{T,V,M}\right.$$
$$+ \left(\frac{\partial \ln Z_G}{\partial \ln \lambda_A}\right)_{T,V,\lambda_B}\left(\frac{\partial \ln \lambda_A}{\partial \bar{N}}\right)_{T,V,\bar{M}}$$
$$\left.+ \left(\frac{\partial \ln Z_G}{\partial \ln \lambda_B}\right)_{T,V,\lambda_A}\left(\frac{\partial \ln \lambda_B}{\partial \bar{N}}\right)_{T,V,\bar{M}}\right]. \tag{12.33}$$

The sum of the terms in brackets is identically zero,[1] so that

$$\left(\frac{\partial \bar{A}}{\partial \bar{N}}\right)_{T,V,\bar{M}} = kT \ln \lambda_A. \tag{12.34}$$

The absence of $\bar{N}$ or $\bar{M}$ from the right-hand side means that the value of $\partial A/\partial N$ for any system in the ensemble is independent of the number of molecules of each kind therein, and *this value is therefore the same for each system within the ensemble irrespective of its material content.*

For any system containing N molecules of A, M molecules of B,..., $(\partial A/\partial N)_{T,V,M}$ is recognised as the chemical potential of component A, so that

$$\mu_A = kT \ln \lambda_A, \tag{12.35}$$

and is the same for all systems within the ensemble, and since when setting up the ensemble of systems we emphasised that each may consist of more than one phase, it follows that μ_A must be the same in each phase.

By precisely the same argument

$$\mu_B = kT \ln \lambda_B, \tag{12.36}$$

and the same remarks apply.

It is evident that the condition for phase equilibrium (that the chemical potential of any component is the same in all parts) has emerged during the analysis of the grand canonical ensemble just as imperceptibly as the condition for thermal equilibrium emerged during the analysis of the canonical ensemble in Section 8.2.

A second conclusion can be drawn which is of equal importance. Suppose that A and B, the two components of the ensemble, can exist both as separate species *and* as a combined molecule AB. The fact that μ_A is the same throughout the ensemble means that the chemical potential of A is the same whether it is present as a separate species or as part of the molecule AB, and the same applies to the chemical potential of B, so that if we care to speak of the chemical potential of the molecule AB, we see that it is the same as the sum of that of A and that of B. In other words, if the reaction

$$A + B \rightleftharpoons AB$$

is possible, the equilibrium condition is that governed by the condition

$$\mu_{A_{eq}} + \mu_{B_{eq}} = \mu_{AB_{eq}}. \tag{12.37}$$

---

[1] See mathematical theorem 8(i) page xvi. Put $C = \ln Z_G$, $X = \ln \lambda_A$, $Y = \ln \lambda_B$, $E = \bar{N}$, $F = \bar{M}$, then impose constancy of T, V on all terms.

The same argument can be extended to cover more complicated reactions, thus if

$$aA + bB \rightleftharpoons cA_{a/c}B_{b/c},$$
$$a\mu_{A_{eq}} + b\mu_{B_{eq}} = c\mu_{A_{a/c}B_{b/c(eq)}}. \qquad (12.38)$$

These conditions may, of course, be expressed in terms of the quantities λ_A, λ_B and λ_{AB}, it following from the general relation $\mu = kT \ln \lambda$ that for the reaction

$$A + B \rightleftharpoons AB$$
$$\lambda_{A_{eq}} \cdot \lambda_{B_{eq}} = \lambda_{AB_{eq}}, \qquad (12.39)$$

and for the reaction

$$aA + bB \rightleftharpoons cA_{a/c}B_{b/c}$$
$$\lambda_{A_{eq}}^a \cdot \lambda_{B_{eq}}^b = \lambda_{A_{a/c}B_{b/c}}^c. \qquad (12.40)$$

We emphasise that these relations hold irrespective of the *nature* of the species, i.e. they are *not* limited to reactions between independent particles as was the equivalent relation established in Section 7.1.

It follows also that when constructing Z_G for a system containing molecules of A and B which can interact to form AB, it is immaterial whether we write

$$Z_G = \sum_{N,M,j} \lambda_A^N \lambda_B^M \exp(-\beta E_{N,M,j})$$

in which we do not choose to distinguish the condition of A and B, or whether we write

$$Z_G = \sum_{N,M,j} \lambda_A^{N-\xi_{eq}} \lambda_B^{M-\xi_{eq}} \lambda_{AB}^{\xi_{eq}} \exp(-\beta E_{N,M,j})$$

where ξ_{eq} is the extent of reaction at equilibrium.

We emphasise lastly that we trace back both the condition for phase equilibrium and the general condition for chemical equilibrium to the fact that when setting up the equations for the equilibrium state of the ensemble, we found it necessary to use only a single multiplier for each component present therein.

The quantities λ_A and λ_B are called the *absolute activities* of components A and B. Those readers familiar with the classical thermodynamics of mixtures will know that the activity of a component as there used is defined *by reference to a selected standard state*. The quantity λ is absolute in the sense that it is independent of the notion of standard states.

The Gibbs function for any system containing N molecules of A and M of B is given by the equation

$$G = N\mu_A + M\mu_B.$$

It follows that the mean value for the Gibbs function for a system in the ensemble is given by the expression

$$\bar{G} = NkT \ln \lambda_A + MkT \ln \lambda_B. \tag{12.41}$$

From classical thermodynamics,

$$G = A + PV.$$

Comparison of (12.32) and (12.41) shows that

$$\bar{P}V = kT \ln Z_G, \tag{12.42}$$

where $\bar{P}$ is the mean pressure of a system in the ensemble.

Before proceeding we draw attention to the fact that since the temperature and values of the chemical potentials are the same throughout the ensemble, the values of λ_A^N and λ_B^M, i.e. $N \exp(\mu_A/kT)$ and $M \exp(\mu_B/kT)$, are necessarily dependent only on N and M and *not* on the energy states j. It follows that we can write equation (12.18) in the form

$$Z_G = \sum_{N,M} \lambda_A^N \lambda_B^M \left[\sum_j \exp(-\beta E_{N,M,j}) \right].$$

Inspection of the quantity in brackets shows that this is the canonical partition function Q, dependent only on T, N and M and summed over all energy states. We can therefore express Z_G more economically by the expression

$$Z_G = \sum_{N,M} \lambda_A^N \lambda_B^M Q_{N,M}. \tag{12.43}$$

12.5   Fluctuations in the Systems of the Grand Canonical Ensemble

We succeeded, in Section 12.3 in obtaining expressions for the ensemble averages $\bar{N}$, $\bar{M}$ and $\bar{E}$. We must now justify the adoption of these values as the most probable for any system in the ensemble. In fact we may accept without further ado that the chance that the energy of a macroscopic system in the ensemble differs perceptibly from $\bar{E}$ is vanishingly small, because this has already been established for the canonical ensemble, and as far as the energy is concerned, the canonical and grand canonical ensembles are equivalent. We can also simplify the exercise somewhat by considering an ensemble containing only a single component, for which, following (12.43) we first write

$$Z_G = \sum_N \lambda^N Q_N,$$

and then, making use of the relation $\lambda = \exp(\mu/kT)$,

$$Z_G = \sum_N Q_N \exp(N\mu/kT). \tag{12.44}$$

It is clear from the study of fluctuations made in Chapter 8, that we require the value of $\overline{\delta N^2}$, and that this is given by the equation

$$\overline{\delta N^2} = \overline{N^2} - (\overline{N})^2. \tag{12.45}$$

Since the number of systems possessing N molecules irrespective of the energy is given by the equation

$$\mathcal{N}_N = \frac{\lambda^N Q_N}{Z_G},$$

and

$$\mathcal{N}_{\overline{N}} = \sum_N \mathcal{N}_N N,$$

$$\overline{N} = \frac{\sum_N N Q_N \exp(N\mu/kT)}{Z_G},$$

so that the expression

$$\overline{N} Z_G = \sum_N N Q_N \exp(N\mu/kT) \tag{12.46}$$

is an identity, as is the expression

$$\overline{N^2} Z_G = \sum_N N^2 Q_N \exp(N\mu/kT). \tag{12.47}$$

We differentiate the left-hand side of (12.46) with respect to μ, keeping V and T constant (this implies constancy of Q_N) to obtain

$$Z_G \left(\frac{\partial \overline{N}}{\partial \mu}\right)_{V,T} + \overline{N}\left(\frac{\partial Z_G}{\partial \mu}\right)_{V,T}$$

$$= Z_G \left(\frac{\partial \overline{N}}{\partial \mu}\right)_{V,T} + \overline{N} \sum_N \left(\frac{\partial Q_N \exp(N\mu/kT)}{\partial \mu}\right)_{V,T}$$

$$= Z_G \left(\frac{\partial \overline{N}}{\partial \mu}\right)_{V,T} + \overline{N} \sum_N \frac{N}{kT} Q_N \exp(N\mu/kT)$$

$$= Z_G \left(\frac{\partial \overline{N}}{\partial \mu}\right)_{V,T} + \frac{(\overline{N})^2}{kT} Z_G.$$

Differentiating the right-hand side of (12.46) we obtain

$$\sum_N \frac{N^2}{kT} Q_N \exp(N\mu/kT),$$

which, from (12.47) equals $\overline{N^2} Z_G/kT$. It follows that

$$Z_G \left(\frac{\partial \overline{N}}{\partial \mu}\right)_{V,T} + \frac{(\overline{N})^2}{kT} Z_G = \frac{\overline{N^2} Z_G}{kT},$$

or, re-arranging and making use of (12.45) that

$$\overline{\delta N^2} = \frac{kT}{(\partial \mu/\partial \overline{N})_{V,T}}. \tag{12.48}$$

This result is independent of the nature of the molecules concerned. We shall evaluate it only for the case of a perfect gas for which

$$\mu = -kT \ln \frac{f}{N}.$$

Since f is a function only of T and V,

$$\left(\frac{\partial \mu}{\partial \overline{N}}\right)_{V,T} = kT \frac{d \ln \overline{N}}{d\overline{N}} = \frac{kT}{\overline{N}},$$

so that

$$\overline{\delta N^2} = \overline{N},$$

and therefore

$$\frac{(\overline{\delta N^2})^{\frac{1}{2}}}{\overline{N}} = \frac{1}{\overline{N}^{\frac{1}{2}}}. \tag{12.49}$$

It is clear that fluctuations in the 'number density' (the number of molecules per unit volume) become appreciable only in the case of a highly attenuated gas. Such fluctuations in the upper atmosphere cause (and are revealed by) the blueness of the sky.

In the case of an open macroscopic system, 'large enough to handle in the laboratory', where N cannot be less than about 10^{20}, we see that

$$\frac{(\overline{\delta N^2})^{\frac{1}{2}}}{\overline{N}} \simeq 10^{-10}.$$

This result means that although in principle, any open macroscopic system in an ensemble may assume any material content, the number of molecules actually assumed deviates so slightly from the mean that it is, for all practical purposes, indistinguishable from it. We are therefore completely justified in regarding the value $\overline{N}$ given by equation (12.23) as the most probable value of N for the open system, and the value which would be obtained were it determined experimentally. What passes for N passes also for the number of molecules of other components in a multi-component system.

12.6   Expressions for the Thermodynamic Properties of an Open Macroscopic System in Terms of the Grand Canonical Partition Function

We now focus attention on one particular system in the ensemble, and consider the relation to it of all others. The only relation that the remainder have to a single system of interest is that they and it can exchange energy in the form of heat, (so that they and it are characterised by the same equilibrium temperature,) and that they and it can exchange particles, (so that the molecules of the same species have the same chemical potential throughout.) The characteristics of the single system of interest are therefore exactly the same whether it is part of the ensemble, or whether all other systems are replaced by reservoirs of molecules of the same species free to pass into or out of the system, and the system and reservoirs are in equilibrium with a heat bath at the same temperature as that of the ensemble.

Objections may of course be raised that all this is of little practical significance because we do not usually study systems 'in equilibrium with reservoirs of molecules of the same species'. We shall dispose of this objection in the next section, but before doing so, point out that in fact we do just this when studying a (necessarily) open phase within a multiphase system. When we make measurements on the vapour phase in equilibrium with a pure liquid or liquid mixture, we are, in effect, studying an open system (the vapour phase) in equilibrium with a reservoir of the same components (the liquid phase), or, if we are more interested in the liquid phase we may regard this as the open system and the vapour phase as a reservoir of particles in equilibrium with it.

It follows therefore that we can equate all properties of an open thermo-statted macroscopic system *or those of an open phase within a closed thermo-statted system* with the ensemble averages established in Sections 12.3 and 12.4, (in the first case, Z_G is the partition function for the system as a whole, and in the second case, it is the partition function for the phase,) and write:

$$N = \left(\frac{\partial \ln Z_G}{\partial \ln \lambda_A} \right)_{T,V,\lambda_B},$$
(12.50)

$$M = \left(\frac{\partial \ln Z_G}{\partial \ln \lambda_B} \right)_{T,V,\lambda_A},$$
(12.51)

$$E = kT^2 \left(\frac{\partial \ln Z_G}{\partial T} \right)_{\lambda_A, \lambda_B, V}, \tag{12.52}$$

$$S = k \ln Z_G - Nk \ln \lambda_A - Mk \ln \lambda_B + kT \left(\frac{\partial \ln Z_G}{\partial T} \right)_{\lambda_A, \lambda_B, V}, \tag{12.53}$$

$$A = NkT \ln \lambda_A + MkT \ln \lambda_B - kT \ln Z_G, \tag{12.54}$$

$$\mu_A = kT \ln \lambda_A, \tag{12.55}$$

$$\mu_B = kT \ln \lambda_B, \tag{12.56}$$

$$G = N\mu_A + M\mu_B = NkT \ln \lambda_A + MkT \ln \lambda_B, \tag{12.57}$$

$$PV = kT \ln Z_G, \tag{12.58}$$

where

$$Z_G = \sum_{N,M,j} \lambda_A^N \lambda_B^M \exp(-E_{N,M,j}/kT)$$

$$= \sum_{N,M} \lambda_A^N \lambda_B^M Q_{N,M}. \tag{12.59}$$

12.7 Thermodynamic Equivalence of Ensembles

We now explain that the use of the expressions given above is by no means limited to open systems and open phases.

(a) The significance of the fact that there is only a vanishingly-small chance that an open macroscopic system maintained at constant temperature and at constant volume acquires numbers of molecules differing significantly from the most probable values, is that the value of $\ln Z_G$ calculated from the expression

$$\ln Z_G = \ln \sum_{N,M} t_{N,M} = \sum_{N,M} \lambda_A^N \lambda_B^M Q_{N,M} \tag{12.60}$$

is indistinguishable from that of the logarithm of the maximum 'term', [1] $\ln t_{N^*, M^*}$ where

$$\ln t_{N^*, M^*} = \ln \lambda_A^{N^*} \lambda_B^{M^*} Q_{N^*, M^*} \tag{12.61}$$

and where the numbers N^* and M^* are those given by equations (12.50) and (12.51).

---

[1] It should perhaps be pointed out that although we have spoken of a maximum 'term', t_{N^*, M^*} encompasses a very large number of quantum states because Q_{N^*, M^*} is summed over all possible energy levels to each of which corresponds $W(E_i)$ quantum states.

(b) Furthermore, when we use the relations of the grand canonical ensemble but substitute into them $\ln t_{N^*, M^*}$ for $\ln Z_G$, we obtain the relations of the canonical ensemble. This is demonstrated by substituting the right-hand side of (12.61) into any of equations (12.52) to (12.54). Thus, choosing (12.53), we have

$$S = k \ln [\lambda_A^{N^*} \lambda_B^{M^*} Q_{N^*, M^*}] - N^* k \ln \lambda_A - M^* k \ln \lambda_B$$

$$+ kT \left(\frac{\partial \ln [\lambda_A^{N^*} \lambda_B^{M^*} Q_{N^*, M^*}]}{\partial T} \right)_{\lambda_A, \lambda_B, V}$$

$$= k \ln Q_{N^*, M^*} + kT \left(\frac{\partial \ln Q_{N^*, M^*}}{\partial T} \right)_V. \qquad (12.62)$$

Comparison with equation (8.47) shows that (12.62) is the canonical ensemble expression for the entropy of a *closed* thermostatted system containing N^* molecules of A and M^* of B.

(c) It will be recalled that it was shown in Section 8.3 that the value of $\ln Q$ for a macroscopic system calculated from the expression

$$Q = \sum_i W(E_i) \exp(-E_i/kT) \qquad (8.49)$$

is indistinguishable from that of the logarithm of the maximum term in the sum, i.e. $\ln W(E^*) - E^*/kT$, where E^* is the most probable value for the energy.

It follows that the value of $\ln Z_G$ calculated from (12.60) is indistinguishable from that of the term

$$\ln [\lambda_A^{N^*} \lambda_B^{M^*} W(E^*) \exp(-E^*/kT)] \qquad (12.63)$$

where N^*, M^* and E^* are the most probable values, and $W(E^*)$ the number of quantum states accessible to an *isolated* system containing N^* molecules of one kind and M^* of another and characterised by energy E^*.

We now find that if we substitute (12.63) for $\ln Z_G$ into any of the relations of the grand canonical ensemble, we recover the relations of the micro-canonical ensemble. This is demonstrated by calculating the entropy of an isolated system of energy E^* containing N^* molecules of A and M^* molecules of B by means of equation (12.53). We then have

$$S = k \ln [\lambda_A^{N^*} \lambda_B^{M^*} W(E^*) \exp(-E^*/kT)]$$

$$- N^* k \ln \lambda_A - M^* k \ln \lambda_B$$

$$+ kT \left(\frac{\partial \ln [\lambda_A^{N^*} \lambda_B^{M^*} W(E^*) \exp(-E^*/kT)]}{\partial T} \right)_{\lambda_A, \lambda_B, V}$$

$$= k \ln W(E^*) - \frac{E^*}{T} + kT\frac{E^*}{kT^2}$$

$$= k \ln W(E^*), \tag{12.64}$$

which is the fundamental relation from which all equations of the micro-canonical ensemble were derived. We can, of course, *evaluate* $W(E^*)$ only for a system of *independent* particles, but the relation

$$S = k \ln W(E^*)$$

holds for all systems.

The significance of (a) is that if when evaluating the properties of an *open* thermostatted system we experience difficulty in summing Z_G we can replace $\ln Z_G$ by the logarithm of its maximum 'term'.

The significance of (b) is that when confronted with a problem concerning either interacting or independent particles, we can regard the system as *closed* (T, V, N, M constant) and use the relations of the canonical ensemble, *or*, if we prefer, regard the system as *open* (T, V, λ_A, λ_B constant) and use the relations of the grand canonical ensemble.

The significance of (c) is that when confronted with a problem concerning independent particles we can regard the system as *isolated* (E, V, N, M constant) and use the relations of the micro-canonical ensemble, *or*, regard it as *closed* (T, V, N, M constant) and use the relations of the canonical ensemble, *or*, regard it as *open* (T, V, λ_A, λ_B constant) and use the relations of the grand canonical ensemble. *In all cases we are free to choose that set of relations which is easiest to handle.*

We have already seen that the relations of the canonical ensemble are simpler to use than those of the micro-canonical ensemble, (for examples, study Exercises 8.3, 9.2 and 10.3) because in the latter we have to take into account the change in temperature resulting from some processes in isolated systems, but in the former we consider the temperature fixed and sum over all values of E. We now remark that the relations of the grand canonical ensemble are often simpler to use than those of the canonical ensemble. The fundamental reason for this is that changes in a closed system may result in changes in the chemical potentials of the components, and that to take these changes into account is often more troublesome than to regard the chemical potentials as fixed, and to sum over all values of N and M.

The greater ease with which the relations of the grand canonical ensemble may be used applies particularly to problems concerning multiphase systems, and even more particularly to those problems in which we are more interested in what is happening in one phase than in what is happening to the system as a whole. The relations of the grand canonical ensemble are such that we are able to regard the phase of particular interest as the 'system', and the other phases merely as reservoirs of components with which the 'system' is in equilibrium. We shall illustrate this in the next chapter by re-considering the important topic of the adsorption of gases, but first, in order to obtain familiarity with the relations of the grand canonical ensemble, we shall re-examine perfect gas systems, and then consider the particular problem encountered in applying the relations of the ensemble to a system of localised particles.

12.8   Re-Examination of Perfect Gas Systems

1. Consider a system of independent non-localised particles of like kind maintained at temperature T and volume V. In Section 8.8 we accepted that for a perfect gas system containing N molecules of the same kind,

$$Q = \frac{f^N}{N!},$$

so that it follows from (12.43) that

$$Z_G = \sum_N \frac{\lambda^N f^N}{N!}.$$

We see immediately[1] that

$$Z_G = \exp(\lambda f),$$

and

$$\ln Z_G = \lambda f. \qquad (12.65)$$

It follows from (12.50) that

$$N = \lambda \left(\frac{\partial \ln Z_G}{\partial \lambda} \right)_{T,V} = \lambda f, \qquad (12.66)$$

from (12.52) that

$$E = kT^2 \left(\frac{\partial \ln Z_G}{\partial T} \right)_{\lambda,V} = kT^2 \lambda \left(\frac{\partial f}{\partial T} \right)_V$$

$$= \frac{NkT^2}{f} \left(\frac{\partial f}{\partial T} \right)_V = NkT^2 \left(\frac{\partial \ln f}{\partial T} \right)_V, \qquad (12.67)$$

[1] $\exp x = x^0 + x + \dfrac{x^2}{2!} + \dfrac{x^3}{3!} + \cdots + \dfrac{x^n}{n!} \cdots .$

from (12.53) that

$$S = k \ln Z_G - Nk \ln \lambda + kT \left(\frac{\partial \ln Z_G}{\partial T} \right)_{\lambda, V}$$

$$= Nk + Nk \ln \frac{f}{N} + NkT \left(\frac{\partial \ln f}{\partial T} \right)_V, \tag{12.68}$$

from (12.54) that

$$A = NkT \ln \lambda - kT \ln Z_G$$

$$= -NkT \ln \frac{f}{N} - NkT, \tag{12.69}$$

from (12.57) that

$$G = NkT \ln \lambda = -NkT \ln \frac{f}{N}, \tag{12.70}$$

and from (12.58) that

$$PV = kT \ln Z_G = kT\lambda f = NkT. \tag{12.71}$$

It will be recognised that these results are the same as those obtained in Chapter 4 using the relations of the micro-canonical ensemble.

2. We now consider a mixture of perfect gases A and B. From (8.85) we have

$$Q_{N,M} = \frac{f_A^N}{N!} \cdot \frac{f_B^M}{M!},$$

so that

$$Z_G = \sum_{N,M} \lambda_A^N \lambda_B^M Q_{N,M}$$

$$= \sum_N \frac{(\lambda_A f_A)^N}{N!} \cdot \sum_M \frac{(\lambda_B f_B)^M}{M!}$$

$$= \exp(\lambda_A f_A) \cdot \exp(\lambda_B f_B)$$

$$= \exp(\lambda_A f_A + \lambda_B f_B),$$

and

$$\ln Z_G = \lambda_A f_A + \lambda_B f_B. \tag{12.72}$$

It follows that

$$N = \lambda_A \left(\frac{\partial \ln Z_G}{\partial \lambda_A} \right)_{T,V,\lambda_B} = \lambda_A f_A, \tag{12.73}$$

$$M = \lambda_B \left(\frac{\partial \ln Z_G}{\partial \lambda_B} \right)_{T,V,\lambda_A} = \lambda_B f_B, \tag{12.74}$$

$$E = kT^2 \left(\frac{\partial \ln Z_G}{\partial T} \right)_{\lambda_A,\lambda_B,V}$$

$$= NkT^2 \left(\frac{\partial \ln f_A}{\partial T} \right)_V + MkT^2 \left(\frac{\partial \ln f_B}{\partial T} \right)_V, \tag{12.75}$$

$$S = k \ln Z_G - Nk \ln \lambda_A - Mk \ln \lambda_B + kT \left(\frac{\partial \ln Z_G}{\partial T} \right)_{\lambda_A,\lambda_B,V}$$

$$= Nk + Mk + Nk \ln \frac{f_A}{N} + Mk \ln \frac{f_B}{M}$$

$$+ NkT \left(\frac{\partial \ln f_A}{\partial T} \right)_V + MkT \left(\frac{\partial \ln f_B}{\partial T} \right)_V, \tag{12.76}$$

$$\vdots$$

$$PV = kT \ln Z_G = kT(N + M). \tag{12.77}$$

It will be recognised that these results are the same as those obtained in Chapter 6.

3. As the last perfect gas system to be re-examined, we choose a system containing two components A and B, but suppose that a reaction

$$aA + bB \rightleftharpoons cC$$

is possible.

We denote the number of molecules of each species by the terms $N - a\xi$, $M - b\xi$, and $c\xi$, where ξ is the extent of reaction (see equation 7.3) and N and M are arbitrary numbers. It follows that

$$Z_G = \sum_{N,M} \frac{(\lambda_A f_A)^{N-a\xi}}{(N - a\xi)!} \cdot \frac{(\lambda_B f_B)^{M-b\xi}}{(M - b\xi)!} \cdot \frac{(\lambda_C f_C)^{c\xi}}{(c\xi)!}$$

$$= \exp{(\lambda_A f_A)} \cdot \exp{(\lambda_B f_B)} \cdot \exp{(\lambda_C f_C)},$$

so that

$$\ln Z_G = \lambda_A f_A + \lambda_B f_B + \lambda_C f_C. \tag{12.78}$$

The equilibrium value for $N - a\xi$ is given by the relation

$$N - a\xi_{eq} = \lambda_A \left(\frac{\partial \ln Z_G}{\partial \lambda_A} \right)_{T,V,\lambda_B,\lambda_C} = \lambda_A f_A,$$

that for $M - b\xi$ by the relation

$$M - b\xi_{eq} = \lambda_B \left(\frac{\partial \ln Z_G}{\partial \lambda_B} \right)_{T,V,\lambda_A,\lambda_C} = \lambda_B f_B$$

and that for $c\xi$ by the relation

$$c\xi_{eq} = \lambda_C \left(\frac{\partial \ln Z_G}{\partial \lambda_C} \right)_{T,V,\lambda_A,\lambda_B} = \lambda_C f_C.$$

Hence

$$\frac{[c\xi_{eq}]^c}{[N - a\xi_{eq}]^a [M - b\xi_{eq}]^b} = \frac{\lambda_{C_{eq}}^c}{\lambda_{A_{eq}}^a \lambda_{B_{eq}}^b} \cdot \frac{f_C^c}{f_A^a f_B^b}.$$

It follows from (12.40) that

$$\frac{\lambda_{C_{eq}}^c}{\lambda_{A_{eq}}^a \lambda_{B_{eq}}^b} = 1,$$

so that

$$\frac{[c\xi_{eq}]^c}{[N - a\xi_{eq}]^a [M - b\xi_{eq}]^b} = \frac{f_C^c}{f_A^a f_B^b}, \tag{12.79}$$

which is, of course equivalent to (7.11).

12.9 The Grand Canonical Ensemble of Systems of Independent Localised Particles

This exercise is instructive because we come up against a physical restraint on the grand partition function.

It follows from equation (8.78) that the canonical partition function for an assembly of N independent localised particles is f^N, so that it appears to follow from equation (12.59) that the grand canonical partition function is given by the expressions

$$Z_G = \sum_N \lambda^N Q_N = \sum_N \lambda^N f^N. \tag{12.80}$$

This expression is, however, incorrect. As was explained in Section 3.2, the energy levels accessible to a localised particle are determined by V/N the volume accessible to each particle, and since in the grand canonical ensemble we sum over all values of N *but keep the volume fixed*, (i.e. regard V and N as independent variables) each value of N must be associated with its own value for f, so that summing for all values of N it would appear that

$$Z_G = 1 + \lambda f_1 + \cdots + (\lambda f_{N-1})^{N-1} + (\lambda f_N)^N$$
$$+ (\lambda f_{N+1})^{N+1} + \cdots \tag{12.81}$$

where f_1 is the particle partition function when $N = 1$, f_{N-1} is the particle partition function when the system contains $N - 1$ particles and f_N is the value of the partition function when the system contains N particles ... and so on.

This sum is intractable, but the situation is relieved by the physical restriction that a close-packed solid occupying volume V *can* contain only N particles where N is given by the equation

$$N = V/\bar{v},$$

where $\bar{v}$ is the molecular volume. In other words, summation over all values of N is 'forbidden' by the characteristics of the system, and the only permissible term in the expression for Z_G is $\lambda^N f^N$, so that

$$\ln Z_G = N \ln \lambda f. \tag{12.82}$$

It now follows that

$$E = kT^2 \left(\frac{\partial \ln Z_G}{\partial T} \right)_{\lambda, V} = NkT^2 \left(\frac{\partial \ln f}{\partial T} \right)_V, \tag{12.83}$$

$$S = k \ln Z_G - Nk \ln \lambda + kT \left(\frac{\partial \ln Z_G}{\partial T} \right)_{\lambda, V}$$

$$= Nk \ln f + NkT \left(\frac{\partial \ln f}{\partial T} \right)_V \tag{12.84}$$

$$A = NkT \ln \lambda - kT \ln Z_G$$

$$= -NkT \ln f, \tag{12.85}$$

which are, of course, the same results as those obtained in Chapter 2.

EXERCISES

12.1   First see Exercise 8.4. Later in the paragraph from which the quotation used in that exercise was taken, Guggenheim goes on to remark "... the problem 'given the number of molecules of a specified kind in a system evaluate their chemical potential' is less tractable than its converse 'given the chemical potential of a molecular species evaluate the number of molecules' ...". Discuss the bearing of this quotation on the subject of this chapter and in particular on the contents of Section 12.7.

CHAPTER 13

The Adsorption of Gases Revisited

As an example of the fact that the relations of the grand canonical ensemble can be applied to any single open phase, we re-examine the adsorption of gases, previously investigated using the methods of the canonical ensemble in Chapter 10. We shall consider only the case of localised adsorption, leaving the corresponding treatment of mobile adsorption as an exercise for the reader.

13.1 The Adsorption of a Single Component

We use the same symbolisation as in Chapter 10, supposing that a surface possesses B equivalent sites on each of which a molecule may be adsorbed, and that M molecules are adsorbed leaving $N - M$ in the gas phase.

In the present treatment we regard the mono-layer as the *system*, and the gas phase merely as the reservoir of molecules with which the mono-layer is in equilibrium. The grand partition function for the mono-layer is given by the expression

$$Z_{G_{(a)}} = \sum_{M=0}^{M=B} \lambda^M Q_{(a)M} = \sum_{M=0}^{M=B} \frac{B!}{M!(B-M)!}(\lambda f_{(a)})^M, \tag{13.1}$$

the value of $Q_{(a)M}$ being that given by equation (10.5).

Objections that in the grand canonical partition function, summation should occur over *all* values of M, and not only over those values lying between $M = 0$ and $M = B$ are valid. We may however, take any one of three points of view. The first is that terminating the sum at the term corresponding to $M = B$ is a restriction imposed on the partition function by the characteristics of the system. The second is that the expression is mathematically meaningless for values of M greater than B. The third is that we know that the value of $\ln Z_G$ is indistinguishable from the logarithm of the maximum term in Z_G, and since the maximum term corresponds to some value of M lying between 0 and B, the contributions made by all terms corresponding to values of M greater than B are negligible.

It follows from the binomial theorem that

$$(1 + x)^B = 1 + Bx + \frac{B(B-1)}{2!}x^2 + \cdots + \frac{B!}{n!(B-n)!}x^n \cdots$$

there being one term in the expansion for each value of n from zero to B, so that it is evident from (13.1) that

$$Z_{G_{(a)}} = [1 + \lambda f_{(a)}]^B,$$

and

$$\ln Z_{G_{(a)}} = B \ln [1 + \lambda f_{(a)}]. \tag{13.2}$$

We obtain the equilibrium value of M by making use of equation (12.50), i.e.

$$M = \lambda \left(\frac{\partial \ln Z_G}{\partial \lambda} \right)_{T,V} = \frac{B \lambda f_{(a)}}{1 + \lambda f_{(a)}}. \tag{13.3}$$

Defining θ as M/B as in Chapter 10, we have immediately

$$\theta = \frac{\lambda f_{(a)}}{1 + \lambda f_{(a)}}. \tag{13.4}$$

We now identify λ by the fact that the mono-layer and gas phase are in equilibrium. Since λ is the same in each phase, its value is that given by the equations

$$kT \ln \lambda = \mu_{(g)} = -kT \ln \frac{f_{(g)}}{N - M},$$

and therefore,

$$\lambda = \frac{N - M}{f_{(g)}} = \frac{N - M}{f'_{(g)}V} = \frac{P}{kT f'_{(g)}},$$

where P is the pressure. We see that

$$\theta = \frac{\dfrac{f_{(a)}}{kT f'_{(g)}} \cdot P}{1 + \dfrac{f_{(a)}}{kT f'_{(g)}} \cdot P} \tag{13.5}$$

which is the same as (10.13).

It is instructive to compare the method used above with that used in Section 10.2. There, we derived an expression for the canonical partition function *for the system as a whole*, (equation 10.7), and picked out the maximum term in the sum by making use of equation (10.8). In the present treatment we constructed the grand partition function for the adsorbed *phase*, found that we were able to sum it, and obtained the equilibrium value of M directly by making use of equation (12.50). The present method is the more elegant, requiring only a certain familiarity with algebraic series in order to recognise that the sum of terms on the right-hand side of (13.1) is the sum of terms of the binomial series.

13.2 Two Molecular Species Competing for the Same Sites

We now capitalise on the facility with which the method described above may be used, by considering the more complicated situation in which two molecular species X and Y compete for the same sites.

We therefore consider the case in which we have B equivalent sites, and suppose that we have in all N molecules of X and K of Y, and that of these, M molecules of X and J molecules of Y are adsorbed. The grand partition function for the adsorbed phase is

$$Z_{G_{(a)}} = \sum_{M=0,J=0}^{M+J=B} \lambda_X^M \lambda_Y^J Q_{M,J}, \tag{13.6}$$

where

$$Q_{M,J} = W_{(c)} \cdot f_X^M \cdot f_Y^J,$$

and we have only to evaluate $W_{(c)}$, the number of ways in which M molecules of one kind and J of another can be arranged on B 'labelled' sites.

If the two molecular species were the same, that number would obviously be

$$\frac{B!}{(M+J)!(B-M-J)!},$$

but since the molecules are of two kinds, each of these distributions may be reached in $(M+J)!/M!J!$ ways, so that

$$W_{(c)} = \frac{B!}{M!J!(B-M-J)!}.$$

It follows that

$$Z_{G_{(a)}} = \sum_{M=0,J=0}^{M+J=B} \frac{B!}{M!J!(B-M-J)!}(\lambda_X f_X)^M \cdot (\lambda_Y f_Y)^J. \tag{13.7}$$

Those readers familiar with the multinomial theorem will recognise the argument of the sum as the general term for the expansion

$$[1 + \lambda_X f_X + \lambda_Y f_Y]^B,$$

so that

$$\ln Z_G = B \ln [1 + \lambda_X f_X + \lambda_Y f_Y]. \tag{13.8}$$

The equilibrium values of M and J are therefore those given by the equations

$$M = \lambda_X \left(\frac{\partial \ln Z_G}{\partial \lambda_X} \right)_{T,V,\lambda_Y} = \frac{B \lambda_X f_X}{1 + \lambda_X f_X + \lambda_Y f_Y}, \tag{13.9}$$

and

$$J = \lambda_Y \left(\frac{\partial \ln Z_G}{\partial \lambda_Y} \right)_{T,V,\lambda_X} = \frac{B \lambda_Y f_Y}{1 + \lambda_X f_X + \lambda_Y f_Y}. \tag{13.10}$$

Defining θ_X as M/B and θ_Y as J/B we have

$$\theta_X = \frac{\lambda_X f_X}{1 + \lambda_X f_X + \lambda_Y f_Y} \quad \text{and} \quad \theta_Y = \frac{\lambda_Y f_Y}{1 + \lambda_X f_X + \lambda_Y f_Y}, \tag{13.11}$$

or

$$\lambda_X f_X = \frac{\theta_X}{1 - \theta_X - \theta_Y} \quad \text{and} \quad \lambda_Y f_Y = \frac{\theta_Y}{1 - \theta_X - \theta_Y}, \tag{13.12}$$

or

$$\frac{\theta_X}{\theta_Y} = \frac{\lambda_X f_X}{\lambda_Y f_Y}. \tag{13.13}$$

We can now identify λ_X and λ_Y by the fact that the adsorbed molecules are in equilibrium with $N - M$ molecules of X and $K - J$ molecules of Y in the gas phase, so that

$$kT \ln \lambda_X = \mu_X = -kT \ln \frac{f_{X_{(g)}}}{N - M},$$

and

$$kT \ln \lambda_Y = \mu_Y = -kT \ln \frac{f_{Y_{(g)}}}{K - J},$$

and

$$\lambda_X = \frac{N - M}{f_{X_{(g)}}} = \frac{N - M}{f'_{X_{(g)}} V} = \frac{P_X}{kT f'_{X_{(g)}}}$$

and

$$\lambda_Y = \frac{K - J}{f_{Y(g)}} = \frac{K - J}{f'_{Y(g)}V} = \frac{P_Y}{kTf'_{Y(g)}},$$

where P_X and P_Y are the partial pressures of X and Y, so that

$$\frac{\theta_X}{\theta_Y} = \frac{f_X}{f'_{X(g)}} \cdot \frac{f'_{Y(g)}}{f_Y} \cdot \frac{P_X}{P_Y}. \tag{13.14}$$

13.3 Atomic Adsorption of Diatomic Molecules

As a last example of the use of these statistical–mechanical relations in the study of localised adsorption, let us suppose that the molecules in the mono-layer and those in the gas phase are the same component but not the same molecular species. A most important example is that of hydrogen which exists as diatomic molecules in the gas phase, but generally as hydrogen atoms in the mono-layer.

From (13.11) we may write

$$\theta_H = \frac{\lambda_H f_H}{1 + \lambda_H f_H + \sum_r \lambda_r f_r} \tag{13.15}$$

where $\sum_r \lambda_r f_r$ represents the sum of terms for all other species present, and from (13.12)

$$\lambda_H f_H = \frac{\theta_H}{1 - \theta_H - \sum_r \theta_r} \tag{13.16}$$

where $\sum_r \theta_r$ represents the sum of the fractions of total number of sites occupied by other species.

We now have equilibrium between adsorbed atoms H and gas molecules H_2, i.e.

$$2H \rightleftharpoons H_2.$$

It follows from (12.40) that

$$\lambda_H^2 = \lambda_{H_2} \tag{13.17}$$

and from the equations of Section 13.1 that

$$\lambda_{H_2} = \frac{P_{H_2}}{kTf'_{H_2}}. \tag{13.18}$$

Substituting (13.17) and (13.18) into (13.15) we find that

$$\theta_H = \frac{f_H\left(\frac{P_{H_2}}{kTf'_{H_2}}\right)^{\frac{1}{2}}}{1 + f_H\left(\frac{P_{H_2}}{kTf'_{H_2}}\right)^{\frac{1}{2}} + \sum_r \lambda_r f_r} \tag{13.19}$$

or, substituting into (13.16) we find that

$$\frac{\theta_H}{1 - \theta_H - \sum_r \theta_r} = f_H \left(\frac{P_{H_2}}{k T f'_{H_2}} \right)^{\frac{1}{2}}. \tag{13.20}$$

Attention is drawn to the fact that the energy zeros implicit in f_H and f'_{H_2} must be the same. This is achieved by choosing as energy zero the potential energy of hydrogen atoms at infinite separation, so that:

(a) f_H contains the factor $\exp(e/kT)$ where e is the energy required to remove an adsorbed hydrogen atom in its lowest energy state, and

(b) f'_{H_2} contains the factor $\exp(D_0/kT)$ where D_0 is the energy required to dissociate the hydrogen molecule, (see equation (6.43) and Section 7.2).

13.4   More Refined Treatments of Localised Adsorption

In the treatments given in the present chapter and in Chapter 10, we assumed that localised adsorbed molecules are independent, and the fact that the Langmuir equations are obeyed in many cases suggests that this is so. Nevertheless, one of the reasons why the Langmuir equations are not universally obeyed at low temperatures (when localised rather than mobile adsorption might be expected) is undoubtedly due to interactions between adsorbed molecules.

Attempts have been made to take into account molecular interactions in a mono-layer by considering the forces between molecules occupying adjacent sites. These attacks follow very closely the statistical–mechanical approach to the problem of imperfect liquid mixtures described in Sections 11.3 and 11.5, and involve an interaction parameter w' similar to that used therein. One of the most interesting features of this approach is that it predicts a critical value for w'/kT above which a 'homogeneous' mono-layer is unstable, and splits into two stable mono-layers analogous to the phenomenon of limited miscibility in a liquid mixture. An introductory account of work in this field is given by Fowler and Guggenheim.[1]

EXERCISES

13.1   Use the relations of the grand canonical ensemble to establish the fundamental equation governing mobile (non-localised) adsorption. Consider the case in which M molecules are adsorbed and $N - M$ free. Show that

$$Z_{G(a)} = \sum_M \lambda^M Q_{(a)M} = \sum_M \frac{(\lambda f_{(a)})^M}{M!}.$$

[1] Fowler and Guggenheim, *Statistical Thermodynamics* (Cambridge University Press, 1949, pp. 429 *et seq.*).

Obtain an expression for the most probable value of M from equation (12.50). Lastly identify λ by reference to the gas phase and show that the most probable value of M is that given by the equation

$$\frac{f(a)}{M} = \frac{f_{(g)}}{N - M}.$$

This equation is the same as (10.18) and can be further developed as shown in Section 10.3.

Fundamental Thermodynamic Formulae

The first law: for any infinitesimal process

$$dE = q + w \tag{1}$$

where q is the heat absorbed by the system and w the work performed on it by the surroundings.

The second law: for any infinitesimal process

$$dS \geqslant \frac{q}{T} \tag{2}$$

where the equality sign holds if the system remains in thermodynamic equilibrium, and the inequality sign holds for any spontaneous (non-equilibrium) process.

Hence for a system of which the energy is determined solely by its entropy and volume (i.e. for a system in which the only form of work possible is expansion work), any infinitesimal process is governed by the single condition

$$T\,dS - dE - P\,dV \geqslant 0. \tag{3}$$

By introducing the auxiliary functions

$$H = E + PV,$$
$$A = E - TS,$$
$$G = E + PV - TS,$$

we obtain three relations which are equivalent to (3) in every sense:

$$T\,dS - dH + V\,dP \geqslant 0 \tag{4}$$
$$-S\,dT - dA - P\,dV \geqslant 0 \tag{5}$$

and

$$-S\,dT - dG + V\,dP \geqslant 0. \tag{6}$$

It is from (5) that we find that for a system in equilibrium

$$\left(\frac{\partial S}{\partial P}\right)_T = -\left(\frac{\partial V}{\partial T}\right)_P$$

which is the Maxwell relation used in Section 5.11.

The (molecular) chemical potential of a component A of a system containing components B, ... is defined by the equations

$$\mu_A = \left(\frac{\partial E}{\partial N_A}\right)_{S,V,N_B,\ldots} = \left(\frac{\partial H}{\partial N_A}\right)_{S,P,N_B,\ldots}$$

$$= \left(\frac{\partial A}{\partial N_A}\right)_{T,V,N_B,\ldots} = \left(\frac{\partial G}{\partial N_A}\right)_{T,P,N_B,\ldots} \tag{7}$$

It follows that any change in a system involving a change in composition is governed by the equations

$$dA = -S\,dT - P\,dV + \sum_{A,B,\ldots} \mu_A\,dN_A, \tag{8}$$

and

$$dG = -S\,dT + V\,dP + \sum_{A,B,\ldots} \mu_A\,dN_A. \tag{9}$$

It follows from (9) that

$$G = \sum_{A,B,\ldots} N_A\,\mu_A, \tag{10}$$

and

$$dG = \sum_{A,B,\ldots} \mu_A\,dN_A + \sum_{A,B,\ldots} N_A\,d\mu_A. \tag{11}$$

Comparison of (9) and (11) shows that

$$-S\,dT + V\,dP = \sum_{A,B,\ldots} N_A\,d\mu_A \tag{12}$$

which is the Gibbs–Duhem equation. In particular, for a two-component system maintained at constant temperature and pressure,

$$N_A\,d\mu_A + N_B\,d\mu_B = 0, \tag{13}$$

or

$$x\,d\mu_A + (1 - x)\,d\mu_B = 0, \tag{14}$$

where x is the mole fraction of component A and $1 - x$ that of B, or

$$x\left(\frac{\partial \mu_A}{\partial x}\right)_{T,P} - (1 - x)\left(\frac{\partial \mu_B}{\partial(1 - x)}\right)_{T,P} = 0. \tag{15}$$

Since in the thermodynamics of liquid mixtures we have the relations

$$\mu_A = \mu_A^\circ + kT \ln \gamma_A x$$

and

$$\mu_B = \mu_B^\circ + kT \ln \gamma_B (1 - x),$$

equation (15) gives

$$x\left(\frac{\partial \ln \gamma_A}{\partial x}\right)_{T,P} - (1 - x)\left(\frac{\partial \ln \gamma_B}{\partial(1 - x)}\right)_{T,P} = 0 \tag{16}$$

which is the form of the Gibbs–Duhem equation used in Section 11.1.

Note

The formulae given above are, as stated earlier, framed for systems for which the only form of work possible is expansion work, so that in proceeding from (1) and (2) to (3) w was replaced by the term $-P\,dV$. The formulae may be generalised to take into account other variables (surface areas, lengths of fibres under tension, and so on), but no case involving additional variables such as these is considered in this book.

Answers to Exercises and Discussion of Some

In most cases the answers to exercises are incorporated in the questions. Where this is not so they are given below, as is the reasoning behind some questions where this is not obvious from the text.

4.1 Since for a system of independent, non-localised particles f is *not* a function of N, it follows from (4.34) that

$$\mu = -kT \ln f/N, \tag{1}$$

but since $P = -(\partial A/\partial V)_T$

$$= NkT \left(\frac{\partial \ln f}{\partial V}\right)_T,$$

$$G = -NkT \ln f/N - NkT + NkTV \left(\frac{\partial \ln f}{\partial V}\right)_T,$$

and so, from the equation $\mu = G/N$

$$\mu = -kT \ln f/N - kT + kTV \left(\frac{\partial \ln f}{\partial V}\right)_T. \tag{2}$$

Equating the right-hand sides of (1) and (2) and dividing by kT we see that

$$\left(\frac{\partial \ln f}{\partial V}\right)_T = \frac{1}{V}.$$

This equation is satisfied only if f is proportional to V.

5.1 (a) $\ln f_t = 70.531$,
(b) $\ln f_r = 4.277$,
(c) $\ln f_e = 1.099$,
(d) $S = 205.1 \ \text{JK}^{-1} \ \text{mol}^{-1}$.

5.4 The symmetry number for N_2 is two, and that of CO unity.

5.5 NO; see page 116

CH$_3$D; each molecule can assume one of four orientations in the lattice

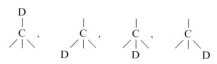

so that for one mole of CH$_3$D at absolute zero,

$$W^0 = 4^L.$$

7.1 (a) $x_{Aeq} = 0.0729$, $x_{Beq} = 0.5365$, $x_{Ceq} = 0.3906$,

(b) $P_{Aeq} = 0.3645$ atm, $P_{Beq} = 2.6825$ atm, $P_{Ceq} = 1.9530$ atm,

(c) $K_P = 2$ atm^{-1},

(d) $x_{Aeq} = 0.056$, $x_{Beq} = 0.472$, $x_{Ceq} = 0.472$,

(e) $P_{Aeq} = 0.280$ atm, $P_{Beq} = 2.360$ atm, $P_{Ceq} = 2.360$ atm,

(f) $K_P = 20$ atm.

7.3 (i) 3.055×10^{-2} atm,

(ii) 8.70×10^{-2}. An experimental value for the degree of dissociation of H$_2$ at this temperature, obtained by Langmuir from his work on the conduction of heat from tungsten filaments in an atmosphere of hydrogen is 7.2×10^{-2}.

7.6 2.47×10^{-2} atm.

7.7 For $P^\dagger = 1$ atm, $K_{P/P^\dagger} = 1.03 \times 10^{13}$, i.e., $K_P = 1.03 \times 10^{13}$ atm,

so that the conversion of acetylene to benzene would be almost complete.

9.1 The energy levels accessible to molecules on the left are governed by V_1 and those on the right by V_2 (see Section 4.2), so that the partition function for particles on the left contains V_1 as a factor, and that for particles on the right contains V_2.

9.3 $V = 8.3 \times 10^{91}$ m^3

9.6 (a) $\Delta H_{subm} = 713.10$ kJ mol^{-1},

(b) $P = 1.62 \times 10^{-112}$ Nm^{-2},

(c) $P = 1.95 \times 10^{-113}$ Nm^{-2},

(d) The value for graphite is 4.93×10^{-113} Nm^{-2}. Since graphite is more stable than diamond at room temperature its vapour pressure must be less than that of diamond. This shows that the value obtained in (b) is to be preferred to that obtained in (c). The reason for the unreliability of (c) is of course that the Einstein treatment gives reasonably accurate results only if θ is very much lower than the temperature concerned, and this condition is not satisfied in (c).

APPENDIX III

Suggested Reading

1. M. Planck, *A Scientific Autobiography* (Williams and Norgate, 1950).

   The first essay in this volume is a fascinating account of the way in which the concept of entropy came into being.

2. G. S. Rushbrooke, *Statistical Mechanics* (Oxford, 1949).

   The approach to some topics (e.g. the derivation of the formula for the W value for an assembly of independent, non-localised particles) differs from that approach chosen in more recent texts. This apart, the book remains an excellent introductory text, its 'feel' for the statistical mechanics of isolated assemblies of independent particles is bettered by none.

3. N. Davidson, *Statistical Mechanics* (McGraw-Hill, 1962).

   A more advanced text which gives an excellent treatment of some topics not covered in this book: (e.g. black body radiation, imperfect gases, dielectric and magnetic phenomena.)

4. T. L. Hill, *An Introduction to Statistical Thermodynamics* (Addison-Wesley, 1960).

   The title is somewhat misleading: few readers would consider it an introductory text. Many chapters are however invaluable to the 'advanced reader'.

5. Sir Ralph Fowler and E. A. Guggenheim, *Statistical Thermodynamics* (Cambridge, 1939; reprinted 1960).

   This is the classical advanced treatise on the applications of statistical mechanics. It is a very valuable book. Unfortunately it is not noted for the ease with which it may be read.

6. E. A. Guggenheim, *Mixtures* (Oxford, 1952).

   An exhaustive account of the development of the quasi-crystalline model of liquid mixtures.

7. E. A. Guggenheim, *Applications of Statistical Mechanics* (Clarendon Press, Oxford, 1965).

   The first chapter gives an invigorating account of the use of the grand canonical ensemble. Later chapters are concerned with developments of the quasi-crystalline model of liquid mixtures since 1952.

8. I. Prigogine, *Molecular Theory of Solutions* (North Holland, 1957).

   An alternative approach to the mechanics of mixtures to that given in 6.

9. T. L. Hill, *Statistical Mechanics* (McGraw-Hill, 1958).

   A more advanced text than 4. A good account of the ensemble for constant P, T and N, and that for constant P, T, μ_A, μ_B (see Section 8.1 of the present book.)

APPENDIX IV

Sources of Thermodynamic Data

F. D. Rossini *et al.*, *Selected Values of Chemical Thermodynamic Properties*, Natl. Bur. Standards Circ. 500, 1952.

F. D. Rossini *et al.*, *Selected Values of Physical and Thermodynamic Properties of Hydrocarbons and Related Compounds*, American Petroleum Institute Research Project 44, Carnegie Press, Pittsburgh, 1953.

Selected Values of Chemical Thermodynamic Properties, Series III, Natl. Bur. Standards 1954.

J.A.N.A.F. Tables of Thermochemical Data, Dow Chemical Company, Midland, Michigan.

D. R. Stull, E. F. Westrum Jr., and G. C. Sinke, *The Chemical Thermodynamics of Organic Compounds* (John Wiley, 1969).

Index

A

Activity, absolute, 270

Activity coefficient (of component of liquid mixture) 234, 236, 237, 248

Adcock, 253

Adsorption of gases, 221 *et seq*, 283 *et seq*
  atomic adsorption of diatomic molecules, 287
  localised, 222, 223 *et seq*
  mobile, 225, 226 *et seq*
  two molecular species competing, 285

Assembly, 21

Atomic crystal, model for, 36
  Debye theory of heat capacity for, 61 *et seq*
  Einstein theory of heat capacity for, 52 *et seq*
  thermodynamic properties of, 56
  vapour pressure of, 213 *et seq*

Azeotropy, 235

B

Barker, 256

Berthelot equation of state, 127

Boltzmann, 26
  constant, 27
  distribution law, 44, 46, 67, 82, 86, 87
  equation, 24

Born–Oppenheimer principle, 93

Bose–Einstein statistics, 78

Broglie, de, equation, 77

C

Canonical ensemble, 180, 183 *et seq*
  partition function, 180, 185
  partition function for assemblies of independent localised particles, 202
  partition function for assemblies of independent non-localised particles, 203

Carbon dioxide, vibrational characteristics of, 111, 128

Carbon monoxide, residual entropy of, 116

Chemical equilibrium, criterion for, 143 *et seq*, 206, 268 *et seq*

Chemical potential, 28
  of component of perfect gas mixture, 138 *et seq*
  of component of liquid mixtures, 231, 233, 245, 248
  of localised particles, 63, 64
  of perfect gas, 124 *et seq*

Clausius, 26

Clausius–Clapeyron equation, 214, 215, 219

Complexion, 10, 21

Component, 20

Configuration, 5, 21, 31

Gibbs function, 121
Grand canonical ensemble, 181, 261 *et seq*
  partition function, 181, 265
  partition function for systems of localised particles, 281
  partition function for systems of non-localised particles, 278
Guggenheim, 182, 183, 240, 252, 254, 282, 288, 297

H

Harmonic oscillator, simple, 53
Heat, statistical-mechanical interpretation of, 131
Henry's law, 257
Helmholtz function, 48
Hill, 182, 183, 297, 298
Hydrogen, residual entropy of, 117 *et seq*

I

Incomplete miscibility, 250 *et seq*
Independent particles, definition of, 36
Indistinguishability of non-localised particles, 82, 84
Internal rotation, 103, 159 *et seq*
  restricted, 160, 163 *et seq*

L

Lagrange method of undetermined multipliers, 42
Langmuir, 221
Lindemann, 59
Liquid mixtures, 231 *et seq*
Localised particles, 36, 202, 281
Longuet–Higgins, 238

M

Maxwell's thermodynamic relations, 127, 292
McGlashan, 253
Mechanical equilibrium, condition for, 29, 200
Micro-canonical ensemble, 183
Mixtures, liquid, 231 *et seq*
  liquid, incomplete miscibility in, 250 *et seq*
  of perfect gases, 132 *et seq*
  perfect liquid, 233, 244 *et seq*
  regular liquid, 237
Moment of inertia, 93 *et seq*
Mulholland, 97

N

Nitric oxide, residual entropy of, 116
Nitrous oxide, residual entropy of, 116
Nuclear spin, 117
  states, 51

O

Ortho and para states, 117

P

Partition function, canonical, 45, 180, 185
  canonical, for system of localised particles, 202
  canonical, for systems of non-localised particles, 203
  canonical, for systems of particles of more than one kind, 204
  canonical, for closed systems containing open phases, 204

Vibration, characteristic temperature of, 104
Table of, 106
Vibrational partition function for localised particle, 55
non-localised particle, 103 *et seq*

Work, statistical-mechanical interpretation of, 131

X

Xylenes, thermodynamics of, 165 *et seq*

W

Z

Water, residual entropy of, 117
Westrum, 124, 299

Zeros, energy, 64, 105, 146, 212, 225
Zeroth law of thermodynamics, 1, 134, 186